revise
STANDARD GRADE
Chemistry

Eileen Ramsden

with Norman Conquest and Tony Buzan

Hodder & Stoughton

A MEMBER OF THE HODDER HEADLINE GROUP

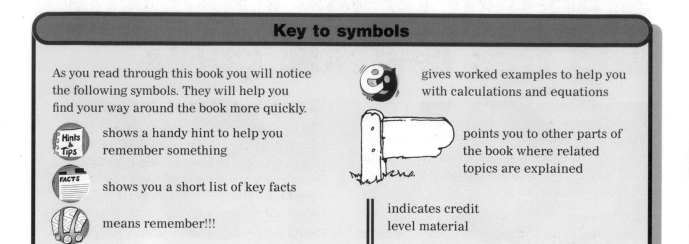

Key to symbols

As you read through this book you will notice the following symbols. They will help you find your way around the book more quickly.

Hints & Tips shows a handy hint to help you remember something

FACTS shows you a short list of key facts

means remember!!!

eg gives worked examples to help you with calculations and equations

points you to other parts of the book where related topics are explained

indicates credit level material

ISBN 0 340 77150 X

First published 2000
Impression number 10 9 8 7 6 5 4 3 2 1
Year 2006 2005 2004 2003 2002 2001 2000

The 'Teach Yourself' name and logo are registered trade marks of Hodder & Stoughton Ltd.

Designed and produced by Gecko Ltd, Bicester, Oxon.
Printed in Spain for Hodder & Stoughton Educational, a division of Hodder Headline Plc,
338 Euston Road, London NW1 3BH, by Graphycems.

Project manager: Jo Kemp
Mind Maps: Patrick Mayfield, Gareth Morris,
  Graham Wheeler
Illustrations: Peter Bull, Simon Cooke, Chris Etheridge,
  Ian Law, Joe Little, Andrea Norton, Mike Parsons,
  John Plumb, Dave Poole, Chris Rothero, Anthony Warne
Cover design: Amanda Hawkes
Cover illustration: Paul Bateman

Standard Grade Chemistry and this revision guide

This revision guide is not intended to replace your textbooks. As tests and examinations approach, however, many students feel the need to revise from something a good deal shorter than their usual textbook. This revision guide is intended to fill that need.

Each revision topic begins with a set of Test Yourself questions to give an idea of how well you have already grasped that topic. You could work through the questions again after you have revised the topic. The improvement should be encouraging! There is a set of Round-up questions at the end of each topic. Work out your Improvement Index from your score on the Round-up questions compared with your first score on the Test Yourself questions.

Organising your time

Make a timetable for homework and revision, and keep to it. You have a lot of subjects to cope with. Leave space in your timetable for your leisure activities. Planned use of time and concentrated study will give you time for your other activities and interests as well as work.

When the exam arrives

The night before the exam make sure that you have everything you will need: your pen and spare cartridge, pencils, rubber, calculator, etc. Decide what you are going to wear and get everything ready. You want to avoid any last minute dithering.

Be optimistic. You have done your revision and can have confidence that it will stand you in good stead. Do not sit up late at night trying to cram. A last-minute glance through the Mind Maps you have made yourself is as much as your brain can take in at the last minute.

In the examination room, read the instructions on the front of the paper before you set pen to paper. Do not spend more time than you should on any one question. If you can't answer a question, move on to the next question and return to the unanswered question later. Attempt all the questions you are supposed to answer. Make sure you turn over every page! Many marks have been lost in exams as a result of turning over two pages at once. If you suffer a panic attack, breathe deeply and slowly to get lots of oxygen into your system and clear your thoughts. Above all, keep your examination in perspective; it is important but not a matter of life or death!

I wish you success.

Eileen Ramsden

Contents

Revision made easy

The four pages that follow contain a gold mine of information on how you can achieve success both at school and in your exams. Read them and apply the information, and you will be able to spend less, but more efficient, time studying, with better results. If you already have another Hodder & Stoughton revision guide, skim-read these pages to remind yourself about the exciting new techniques the books use, then move ahead to page 10.

This section gives you vital information on how to remember more *while* you are learning and how to remember more *after* you have finished studying. It explains

> **how to use special techniques to improve your memory**

> **how to use a revolutionary note-taking technique called Mind Maps that will double your memory and help you to write essays and answer exam questions**

> **how to read everything faster while at the same time improving your comprehension and concentration**

All this information is packed into the next four pages, so make sure you read them!

Your *amazing* memory

There are five important things you must know about your brain and memory to revolutionise your school life.

> **1 how your memory ('recall') works *while* you are learning**

> **2 how your memory works *after* you have finished learning**

> **3 how to use Mind Maps – a special technique for helping you with all aspects of your studies**

> **4 how to increase your reading speed**

> **5 how to zap your revision**

1 Recall during learning – the need for breaks

When you are studying, your memory can concentrate, understand and remember well for between 20 and 45 minutes at a time. Then it *needs* a break. If you carry on for longer than this without one, your memory starts to break down! If you study for hours non-stop, you will remember only a fraction of what you have been trying to learn, and you will have wasted valuable revision time.

So, ideally, *study for less than an hour*, then take a five- to ten-minute break. During the break listen to music, go for a walk, do some exercise, or just daydream. (Daydreaming is a necessary brain-power booster – geniuses do it regularly.) During the break your brain will be sorting out what it has been learning, and you will go back to your books with the new information safely stored and organised in your memory banks. We recommend breaks at regular intervals as you work through the revision guides. Make sure you take them!

2 Recall after learning – the waves of your memory

What do you think begins to happen to your memory straight *after* you have finished learning something? Does it immediately start forgetting? No! Your brain actually *increases* its power and carries on remembering. For a short time after your study session, your brain integrates the information, making a more complete picture of everything it has just learnt. Only then does the rapid decline in memory begin, and as much as 80 per cent of what you have learnt can be forgotten in a day.

However, if you catch the top of the wave of your memory, and briefly review (look back over) what you have been revising at the correct time, the memory is stamped in far more strongly, and stays at the crest of the wave for a much longer time. To maximise your brain's power to remember, take a few minutes and use a Mind Map to review what you have learnt at the end of a day. Then review it at the end of a week, again at the end of a month, and finally a week before the exams. That way you'll ride your memory wave all the way to your exam – and beyond!

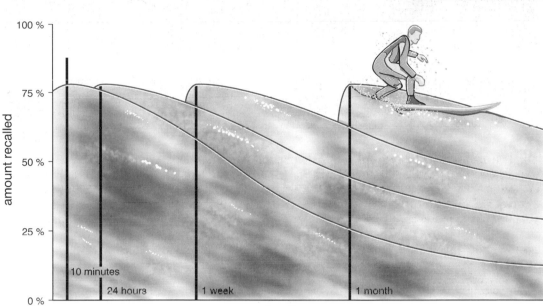

Amazing as your memory is (think of everything you actually do have stored in your brain at this moment) the principles on which it operates are very simple: your brain will remember if it (a) has an image (a picture or a symbol); (b) has that image fixed and (c) can link that image to something else.

3 The Mind Map® – a picture of the way you think

Do you *like* taking notes? More importantly, do you like having to go back over and learn them before exams? Most students I know certainly do not! And how do you take your notes? Most people take notes on lined paper, using blue or black ink. The result, visually, is *boring*! And what does your brain do when it is bored? It turns off, tunes out, and goes to sleep! Add a dash of colour, rhythm, imagination, and the whole note-taking process becomes much more fun, uses more of your brain's abilities, *and* improves your recall and understanding.

A Mind Map mirrors the way your brain works. It can be used for note-taking from books or in class, for reviewing what you have just studied, for revising, and for essay planning for coursework and in exams. It uses all your memory's natural techniques to build up your rapidly growing 'memory muscle'.

You will find Mind Maps throughout this book. Study them, add some colour, personalise them, and then have a go at drawing your own – you'll remember them far better! Put them on your walls and in your files for a quick-and-easy review of the topic.

How to draw a Mind Map

❶ Start in the middle of the page with the page turned sideways. This gives your brain the maximum room for its thoughts.

❷ Always start by drawing a small picture or symbol. Why? Because a picture is worth a thousand words to your brain. And try to use at least three colours, as colour helps your memory even more.

❸ Let your thoughts flow, and write or draw your ideas on coloured branching lines connected to your central image. These key symbols and words are the headings for your topic. The Mind Map at the top of the next page shows you how to start.

❹ Then add facts and ideas by drawing more, smaller, branches on to the appropriate main branches, just like a tree.

❺ Always print your word clearly on its line. Use only one word per line. The Mind Map at the foot of the

next page shows you how to do this.

❻ To link ideas and thoughts on different branches, use arrows, colours, underlining, and boxes.

How to read a Mind Map

❶ Begin in the centre, the focus of your topic.

❷ The words/images attached to the centre are like chapter headings, read them next.

❸ Always read out from the centre, in every direction (even on the left-hand side, where you will have to read from right to left, instead of the usual left to right).

Using Mind Maps

Mind Maps are a versatile tool – use them for taking notes in class or from books, for solving problems, for brainstorming with friends, and for reviewing and revising for exams – their uses are endless! You will find them invaluable for planning essays for coursework and exams. Number your main branches in the order in which you want to use them and off you go – the main headings for your essay are done and all your ideas are logically organised!

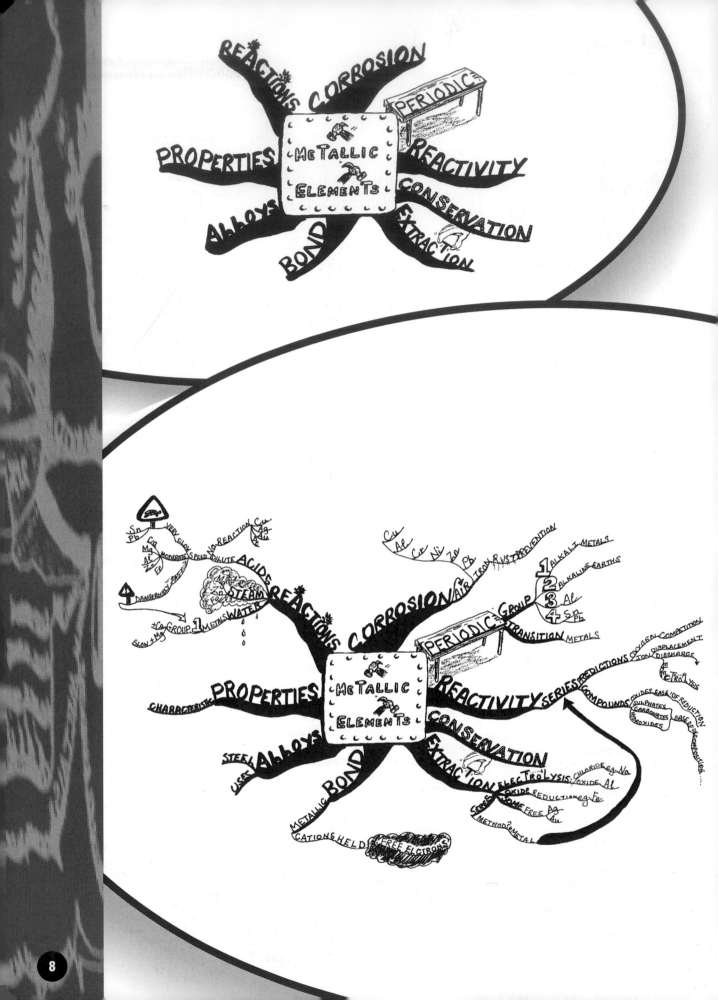

4 Super speed reading

It seems incredible, but it's been proved – the faster you read, the more you understand and remember! So here are some tips to help you to practise reading faster – you'll cover the ground more quickly, remember more, *and* have more time for revision!

★ First read the whole text (whether it's a lengthy book or an exam paper) very quickly, to give your brain an overall idea of what's ahead and get it working. (It's like sending out a scout to look at the territory you have to cover – it's much easier when you know what to expect!) Then read the text again for more detailed information.

★ Have the text a reasonable distance away from your eyes. In this way your eye/brain system will be able to see more at a glance, and will naturally begin to read faster.

★ Take in groups of words at a time. Rather than reading 'slowly and carefully' read faster, more enthusiastically. Your comprehension will rocket!

★ Take in phrases rather than single words while you read.

★ Use a guide. Your eyes are designed to follow movement, so a thin pencil underneath the lines you are reading, moved smoothly along, will 'pull' your eyes to faster speeds.

5 Helpful hints for exam revision

Start to revise at the beginning of the course. Cram at the start, not the end and avoid 'exam panic'!

Use Mind Maps throughout your course, and build a Master Mind Map for each subject – a giant Mind Map that summarises everything you know about the subject.

Use memory techniques such as mnemonics (verses or systems for remembering things like dates and events, or lists).

Get together with one or two friends to revise, compare Mind Maps, and discuss topics.

And finally...

★ *Have fun while you learn* – studies show that those people who enjoy what they are doing understand and remember it more, and generally do it better.

★ *Use your teachers* as resource centres. Ask them for help with specific topics and with more general advice on how you can improve your all-round performance.

★ *Personalise your* revision guide by underlining and highlighting, by adding notes and pictures. Allow your brain to have a conversation with it!

Your brain is an amazing piece of equipment – learn to use it, and you, like thousands of students before you will be able to master exams with ease. The more you understand and use your brain, the more it will repay you!

Matter and chemical reactions

1

preview

At the end of this topic you will be able to:

- **describe the states of matter and changes of state**
- **apply the kinetic theory of matter to solids, liquids, gases, changes of state, dissolving, diffusion and Brownian motion.**
- **describe the structures of some elements**
- **distinguish between an element, a compound and a mixture**
- **write an equation for a chemical reaction.**

How much do you already know? Work out your score on pages 108.

Test yourself

1 Name the three chief states of matter. [3]

2 What can you tell about the purity of a solid from its melting point? [2]

3 What is the difference between evaporation and boiling? [2]

4 How can you tell when a liquid is boiling? [1]

5 Which of the following are chemical reactions?
a) frying an egg
b) boiling a kettle
c) burning a candle
d) melting ice
e) butter turning rancid. [3]

6 Which of the following are compounds (and not mixtures)? Table salt, whisky, wine, water, glucose, margarine, vinegar. [3]

7 Name the elements present in **(i)** nickel sulphide **(ii)** nickel sulphate. [5]

8 Why are crystals shiny? [3]

9 What happens to the heat energy that is supplied to a solid to make it melt? [2]

1.1 States of matter

Everything in the Universe is composed of matter. Matter exists in three chief states: the solid, liquid and gaseous states.

	volume	shape	effect of rise in temperature
solid	fixed	definite	expands slightly
liquid	fixed	flows – changes shape to fit the shape of the container	expands
gas	changes to fit the container	changes to fit the container	expands greatly (gases have much lower densities than solids and liquids)

Characteristics of the solid, liquid and gaseous states

1.2 Change of state

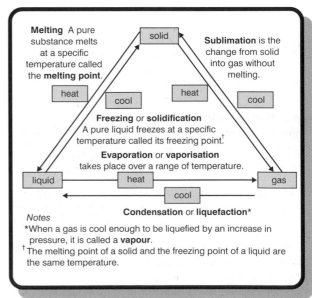

Melting A pure substance melts at a specific temperature called the **melting point**.

Sublimation is the change from solid into gas without melting.

Freezing or **solidification** A pure liquid freezes at a specific temperature called its freezing point.[†]

Evaporation or **vaporisation** takes place over a range of temperature.

Condensation or **liquefaction***

Notes
*When a gas is cool enough to be liquefied by an increase in pressure, it is called a **vapour**.
[†] The melting point of a solid and the freezing point of a liquid are the same temperature.

Matter can change from one state to another

10

1.3 Some properties of materials

★ **Density:** $\text{density} = \dfrac{\text{mass}}{\text{volume}}$.

★ **Melting point:** while a pure solid melts, the temperature remains constant at the melting point of the solid.

★ **Boiling point:** while a pure liquid boils, the temperature remains constant at the boiling point of the liquid.

★ **Conductivity** (thermal and electrical): the ability to conduct heat and electricity is a characteristic of metals and alloys.

★ **Solubility:** a solution consists of a **solute** dissolved in a **solvent**. A concentrated solution contains a high proportion of solute; a dilute solution contains a low proportion of solute. A saturated solution contains as much solute as it is possible to dissolve at the stated temperature.

Solubility is the mass of solute that will dissolve in 100 g of solvent at the stated temperature.

Two ways of expressing concentration are:

$$\text{concentration} = \frac{\text{mass of solute}}{\text{volume of solution}}$$

$$\text{concentration} = \frac{\text{amount (moles) of solute}}{\text{volume of solution}}$$

(see page 17)

1.4 The kinetic theory

According to the **kinetic theory of matter**, all forms of matter are made up of small particles which are in constant motion. The theory explains the states of matter and changes of state.

In a solid, the particles are close together and attract one another strongly. They are arranged in a regular three-dimensional structure. The particles can vibrate, but they cannot move out of their positions in the structure.

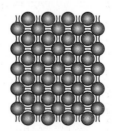
The arrangement of particles in a solid

When the solid is heated, the particles vibrate more energetically. If they gain enough energy, they may break away from the structure and become free to move independently. When this happens, the solid has melted.

In a liquid, the particles are further apart than in a solid. They are free to move about. This is why a liquid flows easily and has no fixed shape. There are forces of attraction between particles. When a liquid is heated, some particles gain enough energy to break away from the other particles and become a gas.

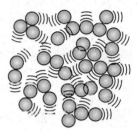
The arrangement of particles in a liquid

Most of a gas is space, through which the particles move at high speed. There are only very small forces of attraction between the particles. When a mass of liquid vaporises, it forms a very much larger volume of gas because the particles are so much further apart in a gas.

Collisions between the gas particles and the container create pressure on the container.

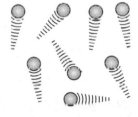
The arrangement of particles in a gas

Crystals

A crystal is a piece of matter with a regular shape and smooth surfaces which reflect light. Viewed through an electron microscope, crystals can be seen to consist of a regular arrangement of particles. The regular arrangement of particles gives the crystal its regular shape.

1.5 What does the kinetic theory explain?

Dissolving of a solid

When a solid dissolves, particles of solid separate from the crystal and spread out through the solvent to form a solution.

Diffusion of a gas

When a gas is released into a container, particles of gas move through the container until the gas has spread evenly through all the space available.

Evaporation or vaporisation

Attractive forces exist between the particles in a liquid. Some particles with more energy than the average break away from the attraction of other particles and escape into the vapour phase. The average energy of the particles that remain is lower than before – the liquid has cooled.

Brownian motion

The botanist William Brown used his microscope a century ago to observe grains of pollen suspended in water. He saw that the grains were in constant motion. The explanation is that water molecules collide with a pollen grain and give it a push. The direction of the push changes as different numbers of molecules strike the pollen grain from different sides.

1.6 Metallic and non-metallic elements

Elements are pure substances that cannot be split up into simpler substances. Some elements exist as **allotropes** – forms of the same element which have different crystalline structures. Allotropes of carbon are shown on page 13. Elements are classified as metallic and non-metallic (see Topic 3).

1.7 Structures of elements

Individual molecules

Some elements consist of small individual molecules with negligible forces of attraction between them, e.g. oxygen O_2 and chlorine Cl_2.

Molecular structures

Some elements consist of molecules held in a crystal structure by weak intermolecular forces. Solid iodine is a structure composed of I_2 molecules; iodine vapour consists of individual I_2 molecules.

Giant molecules

Some elements consist of giant molecules or macromolecules, which are composed of millions of atoms bonded together in a three-dimensional structure, e.g. the allotropes of carbon – diamond, graphite and fullerenes – shown on page 13.

1.8 Chemical reactions

A chemical reaction is a change in which a new substance is formed. The result of a chemical reaction between elements is the formation of a compound, e.g.

magnesium + oxygen → magnesium oxide

calcium → calcium + carbon
carbonate   oxide   dioxide

1.9 Compounds

A **compound** is a pure substance that consists of two or more elements which are chemically combined in fixed proportions by mass. Some compounds can be **synthesised** from their elements, e.g. calcium burns in oxygen to form calcium oxide; hot copper combines with chlorine to form copper chloride.

It may be possible to split up a compound into its elements

- by **thermal decomposition**, e.g. silver oxide splits up into silver and oxygen when heated
- by **electrolysis**, e.g. water is electrolysed to hydrogen and oxygen.

A compound with the name ending in -ide, e.g. magnesium chloride, consists of two elements. A name ending in -ite, e.g. sodium sulphite, or -ate, e.g. copper nitrate, shows that oxygen and two other elements are present.

A compound differs from a mixture of elements as shown in the table on the opposite page.

1.10 Symbols and formulae

Every element has its own **symbol**. The symbol is a letter or two letters which stand for one atom of the element, e.g. aluminium Al, iron Fe.

Every compound has a **formula**. This is composed of the symbols of the elements present along with numbers which give the ratio in which the atoms are present.

A molecule of sulphuric acid (see below) contains 2 hydrogen atoms, 1 sulphur atom and 4 oxygen atoms, giving the formula H_2SO_4.

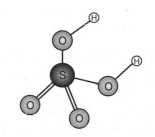

H_2SO_4 – a single molecule

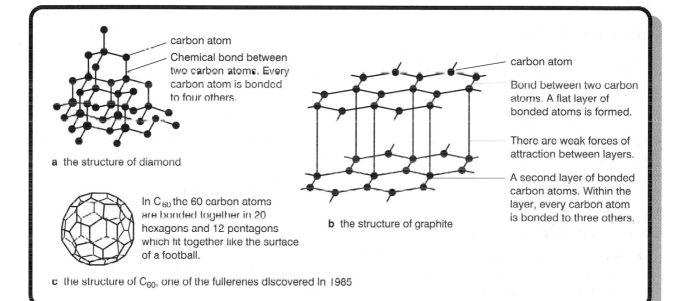

a the structure of diamond

carbon atom

Chemical bond between two carbon atoms. Every carbon atom is bonded to four others.

In C_{60} the 60 carbon atoms are bonded together in 20 hexagons and 12 pentagons which fit together like the surface of a football.

c the structure of C_{60}, one of the fullerenes discovered in 1985

carbon atom

Bond between two carbon atoms. A flat layer of bonded atoms is formed.

There are weak forces of attraction between layers.

A second layer of bonded carbon atoms. Within the layer, every carbon atom is bonded to three others.

b the structure of graphite

The allotropes of carbon

mixtures	compounds
No chemical change takes place when a mixture is made.	When a compound is made, a chemical reaction takes place, and heat is often taken in or given out.
A mixture has the same properties as its components.	A compound has a new set of properties; it does not behave in the same way as the components.
A mixture can be separated into its parts by methods such as distillation (see pages 16–18).	A compound can be split into its elements or into simpler compounds only by a chemical reaction.
A mixture can contain its components in any proportions.	A compound contains its elements in fixed proportions by mass, e.g. magnesium oxide always contains 60% by mass of magnesium.

Differences between mixtures and compounds

Silicon(IV) oxide, shown here, consists of macromolecules which contain twice as many oxygen atoms as silicon atoms, giving the formula SiO_2.

The formula of ammonium sulphate is $(NH_4)_2SO_4$. The '2' multiplies the symbols in brackets: there are 2 nitrogen, 8 hydrogen, 1 sulphur and 4 oxygen atoms.

Writing $2Al_2O_3$ means that the numbers below the line each multiply the symbols in front of them, and the 2 on the line multiplies everything that comes after it, giving a total of 4 aluminium and 6 oxygen atoms.

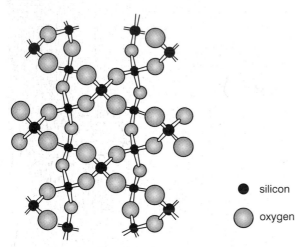

● silicon

○ oxygen

SiO_2 – a macromolecule

1.11 Equations

To write an equation for a chemical reaction:

1 Write a word equation for the reaction.

2 Put in the symbols for the elements and the formulas for the compounds.

3 Put in the **state symbols** (s) for solid, (l) for liquid, (g) for gas, (aq) for in aqueous solution (in water).

Example

1 magnesium oxide + water → magnesium hydroxide solution

2 $MgO + H_2O \rightarrow Mg(OH)_2$

3 $MgO(s) + H_2O(l) \rightarrow Mg(OH)_2(aq)$

Ionic equations

In **ionic equations**, only the ions that take part in the reaction are shown. Here are some examples.

★ **Neutralisation**

acid + alkali → salt + water
$H^+(aq) + OH^-(aq) \rightarrow$ $H_2O(l)$

★ **Displacement**

zinc + copper(II) sulphate solution
  → zinc sulphate solution + copper
$Zn(s) + Cu^{2+}(aq) \rightarrow Zn^{2+}(aq) + Cu(s)$

★ **Precipitation**

barium chloride + sodium sulphate
    solution   solution
  → barium sulphate + sodium chloride
     precipitate   solution
$Ba^{2+}(aq) + SO_4^{2-}(aq) \rightarrow BaSO_4(s)$

round-up

How much have you improved?
Work out your improvement index on page 108.

1 The graph shows temperature against time as a liquid is heated. What is happening at A and B? [3]

2 The graph below shows temperature of a solid against time as it is heated. What is happening at C, D and E? Is the solid a pure substance? [7]

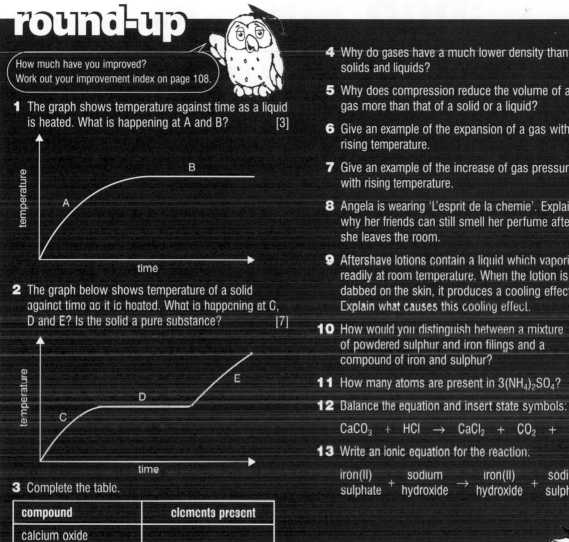

3 Complete the table.

compound	elements present
calcium oxide	
magnesium chloride	
copper(II) sulphide	
copper(II) sulphate	
potassium carbonate	

[12]

4 Why do gases have a much lower density than solids and liquids? [1]

5 Why does compression reduce the volume of a gas more than that of a solid or a liquid? [2]

6 Give an example of the expansion of a gas with rising temperature. [1]

7 Give an example of the increase of gas pressure with rising temperature. [1]

8 Angela is wearing 'L'esprit de la chemie'. Explain why her friends can still smell her perfume after she leaves the room. [3]

9 Aftershave lotions contain a liquid which vaporises readily at room temperature. When the lotion is dabbed on the skin, it produces a cooling effect. Explain what causes this cooling effect. [2]

10 How would you distinguish between a mixture of powdered sulphur and iron filings and a compound of iron and sulphur? [4]

11 How many atoms are present in $3(NH_4)_2SO_4$? [1]

12 Balance the equation and insert state symbols: [4]

$CaCO_3 + HCl \rightarrow CaCl_2 + CO_2 + H_2O$

13 Write an ionic equation for the reaction: [3]

iron(II) sulphate + sodium hydroxide → iron(II) hydroxide + sodium sulphate

Well done if you've improved. Don't worry if you haven't. Take a break and try again.

Reaction speeds

2

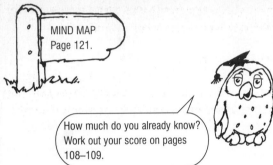

preview

At the end of this topic you will:

- **understand the factors which can change the speed of a chemical reaction.**

MIND MAP
Page 121.

How much do you already know? Work out your score on pages 108–109.

Test yourself

1 Which act faster to cure acid indigestion, indigestion tablets or indigestion powders? Explain your answer. [2]

2 a) Suggest three ways in which you could speed up the reaction between zinc and dilute sulphuric acid:

Zinc + sulphuric acid → hydrogen + zinc sulphate

$Zn(s) + H_2SO_4(aq) \rightarrow H_2(g) + ZnSO_4(aq)$ [3]

b) Explain why each of these methods increases the speed of the reaction. [4]

3 Sketch an apparatus in which you could collect a gaseous product of a reaction and measure the rate at which it was formed. [5]

4 What is a catalyst? [2]

5 Why are catalysts important in industry? [2]

6 Name two reactions which depend on the absorption of light energy. [2]

2.1 Particle size

The reaction between a solid and a liquid is speeded up by using smaller particles of the solid reactant. The reason is that it is the atoms or ions at the surface of the solid that react, and the ratio of surface area:mass is greater for small particles than for large particles.

The diagram below shows an apparatus which you may have used to investigate the effect of particle size on the reaction:

$$\text{calcium carbonate} + \text{hydrochloric acid} \rightarrow \text{carbon dioxide} + \text{calcium chloride} + \text{water}$$

cotton wool stops spray from escaping

dilute hydrochloric acid in a conical flask

calcium carbonate (marble chips)

top–pan balance

The effect of particle size on the speed of a reaction

As the reaction happens, carbon dioxide is given off and the mass of the reacting mixture decreases.

1 Note the mass of flask + acid + marble chips.

2 Add the marble chips to the acid, and start a stopwatch.

3 Note the mass after 10 seconds and then every 30 seconds for 5–10 minutes.

4 Plot the mass against time since the start of the reaction.

5 Repeat with the same mass of smaller chips.

2.2 Concentration

A precipitate of sulphur is formed in the reaction:

| sodium thiosulphate | + | hydrochloric acid | → | sulphur | + | sulphur dioxide | + | sodium chloride | + water |

$$Na_2S_2O_3(aq) + 2HCl(aq) \rightarrow S(s) + SO_2(g) + 2NaCl(aq) + H_2O(l)$$

1   Watch the precipitate of sulphur appear.

2   Note the time when the precipitate is thick enough to block your view of a cross on a piece of paper.

3   Repeat for various concentrations of acid and for various concentrations of thiosulphate.

The experiment shows that, for this reaction,

- rate of reaction is proportional to concentration of thiosulphate
- rate of reaction is proportional to concentration of acid.

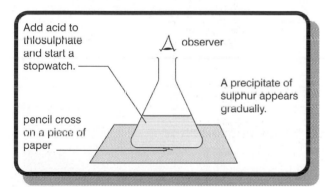

The effect of concentration on the speed of a reaction

2.3 Pressure

An increase in pressure increases the rates of reactions between gases. As the molecules are pushed more closely together, they react more rapidly.

2.4 Temperature

The reaction between thiosulphate and acid can be used to study the effect of temperature on the rate of a reaction, as shown in the following graphs. This reaction goes twice as fast at 30 °C as it does at 20 °C. At higher temperatures, ions have more kinetic energy and collide more often and more vigorously, giving them a greater chance of reacting.

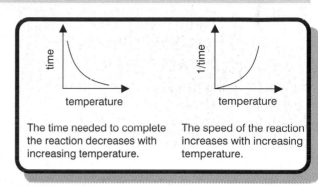

The time needed to complete the reaction decreases with increasing temperature.

The speed of the reaction increases with increasing temperature.

The effect of temperature on the speed of a reaction

2.5 Light

Heat is not the only form of energy that speeds up chemical reactions. Light energy enables many reactions to take place, e.g. photosynthesis and photography.

2.6 Catalysts

Hydrogen peroxide decomposes to form oxygen and water:

hydrogen peroxide → oxygen + water

$$2H_2O_2(aq) \rightarrow O_2(g) + 2H_2O(l)$$

The decomposition takes place very slowly unless a **catalyst**, e.g. manganese(IV) oxide, is present. The rate at which the reaction takes place can be found by collecting the oxygen formed and measuring its volume at certain times after the start of the reaction, as shown in the diagram.

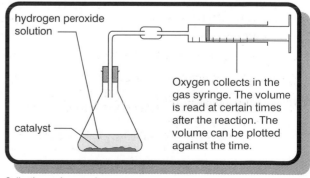

Oxygen collects in the gas syringe. The volume is read at certain times after the reaction. The volume can be plotted against the time.

Collecting and measuring a gas

★ A catalyst is a substance which increases the rate of a chemical reaction without being used up in the reaction.

★ A catalyst will catalyse a certain reaction or group of reactions. Platinum catalyses certain oxidation reactions, and nickel catalyses some hydrogenation reactions.

★ Catalysts are very important in industry. They enable a manufacturer to make a product more rapidly or at a lower temperature.

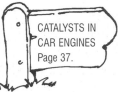

CATALYSTS IN CAR ENGINES Page 37.

2.7 Enzymes

Chemical reactions take place in the cells of living things. These reactions happen at reasonably fast rates at the temperatures which exist in plants and animals. They can do this because the cells contain powerful catalysts called **enzymes**.

Enzymes are proteins. They have large molecules which are twisted into complicated three-dimensional structures. The structures may be damaged by temperatures above about 45°C. Here are some examples of enzyme-catalysed reactions:

- Enzymes in yeast catalyse the conversion of sugar into ethanol and carbon dioxide. The process is called **fermentation**. It is used to make ethanol (alcohol) by the fermentation of carbohydrates. It also produces bubbles of carbon dioxide which make bread rise.

FERMENTATION Page 106.

- Enzymes in bacteria produce yoghurt from milk. They catalyse the conversion of lactose, the sugar in milk, into lactic acid.

round-up

How much have you improved?
Work out your improvement index on page 109.

1 Iron reacts with dilute sulphuric acid to give hydrogen. Which of the following will give **(i)** the slowest reaction **(ii)** the fastest reaction? In each case the mass of iron and the same volume of acid are used.
a) Iron nails + 10% acid at 20°C.
b) Iron nails + 20% acid at 20°C.
c) Iron nails + 20% acid at 30°C.
d) Iron filings + 10% acid at 20°C.
e) Iron filings + 20% acid at 20°C.
f) Iron filings + 20% acid at 30°C. [2]

2 The reaction between calcium carbonate and hydrochloric acid was carried out under the conditions shown in the table.

calcium carbonate	hydrochloric acid/mol/l	temperature/°C
A lumps	1	20
B powder	1	20
C lumps	2	20
D powder	2	20
E lumps	2	30
F powder	2	30

a) Which conditions gave the fastest reaction? [1]
b) Which pairs or pair of experiments can be used to compare the effect of particle size on rate? [3]
c) Which pair or pairs of experiments can be used to compare the effect of temperature on rate? [2]

3 The graphs show the results of experiments on the rate of formation of hydrogen from zinc and an excess of dilute hydrochloric acid at 20°C. Match each of the graphs with the one of the descriptions on page 19.

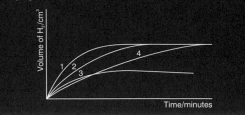

A 1 g zinc powder + 1 mol/1 hydrochloric acid
B 1 g zinc powder + 2 mol/1 hydrochloric acid
C 0.5 g zinc powder + 2 mol/1 hydrochloric acid
D 1 g zinc powder + 0.5 mol/1 hydrochloric acid

[4]

4 The three graphs were obtained in experiments as described on page 16.
   a) Why is there a decrease in mass? [1]
   b) Which of the graphs relates to **(i)** small chips
      (ii) large chips **(iii)** medium-sized chips? [2]
   c) Explain why there is a difference. [1]

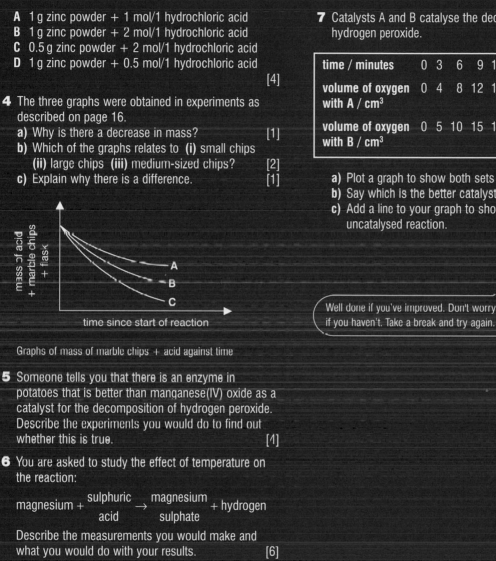

Graphs of mass of marble chips + acid against time

5 Someone tells you that there is an enzyme in potatoes that is better than manganese(IV) oxide as a catalyst for the decomposition of hydrogen peroxide. Describe the experiments you would do to find out whether this is true. [1]

6 You are asked to study the effect of temperature on the reaction:

$$\text{magnesium} + \frac{\text{sulphuric}}{\text{acid}} \rightarrow \frac{\text{magnesium}}{\text{sulphate}} + \text{hydrogen}$$

Describe the measurements you would make and what you would do with your results. [6]

7 Catalysts A and B catalyse the decomposition of hydrogen peroxide.

time / minutes	0	3	6	9	12	15	18	21
volume of oxygen with A / cm³	0	4	8	12	16	17	18	18
volume of oxygen with B / cm³	0	5	10	15	16.5	18	18	18

   a) Plot a graph to show both sets of results. [4]
   b) Say which is the better catalyst, A or B. [1]
   c) Add a line to your graph to show the uncatalysed reaction. [1]

Well done if you've improved. Don't worry if you haven't. Take a break and try again.

Atoms and the periodic table

3

preview

At the end of this topic you will:

- **understand the structure of the periodic table**
- **know the nature of the elements in Groups 0, 1, 2, 7 and the transition elements**
- **know the differences between metallic and non-metallic elements**
- **know the names of the particles of which atoms are composed**
- **know how particles are arranged in the atom**
- **understand the terms atomic number, mass number, relative atomic mass, relative formula mass and isotope.**

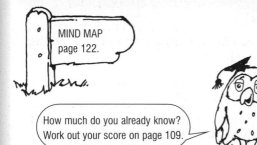

MIND MAP page 122.

How much do you already know? Work out your score on page 109.

Test yourself

1 **a)** What are the noble gases? [2]
   b) In which group of the periodic table are they? [1]
   c) What do the noble gases have in common regarding
      (i) their electron arrangements and
      (ii) their chemical reactions? [2]

2 X is a metallic element. It reacts slowly with water to give a strongly alkaline solution. In which group of the periodic table would you place X? [1]

3 Y is a non-metallic element. It reacts vigorously with sodium to give a salt of formula NaY. In which group of the periodic table would you place Y? [1]

4 Z is a metallic element which reacts rapidly with water to give a flammable gas and an alkaline solution. In which group of the periodic table would you place Z? [1]

5 **a)** Name the halogens. [4]
   b) In which group of the periodic table are they? [1]
   c) Does the chemical reactivity of the halogens increase or decrease with atomic number? [1]
   d) Give the formulas of the products of the reactions of **(i)** sodium **(ii)** iron with each of the halogens. [8]

6 What is a transition metal? Name two transition metals. [2]

7 An atom is made of charged particles called protons and electrons. Why is an atom uncharged? [2]

8 An atom of potassium has mass number 39 and atomic number 19. What is **a)** the number of electrons and **b)** the number of neutrons? [2]

9 Why do the isotopes of an element have the same chemical reactions? [2]

10 What is meant by **a)** the atomic number and **b)** the mass number of an element? [3]

11 Write the symbol, with mass number and atomic number, for each of the following isotopes:
   a) phosphorus with atomic number 15 and mass number 31 [2]
   b) potassium with atomic number 19 and mass number 39. [2]

12 An atom of carbon has 6 electrons. Say how the electrons are divided between shells. [2]

3.1 The periodic table

A major advance in classifying elements was made by John Newlands in 1866 and Dmitri Mendeleev in 1871 when they originated the periodic table. The modern periodic table arranges the elements in order of increasing atomic number.

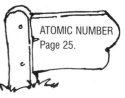

ATOMIC NUMBER Page 25.

A vertical column of elements is a **group** and a horizontal row is a **period**.

The following patterns can be seen in the arrangement of the elements in the periodic table.

1 The reactive metals are at the left-hand side of the table, less reactive metallic elements in the middle block and non-metallic elements at the right-hand side.

2 The differences between the metals in Group 1, those in Group 2 and the transition metals are summarised in the table at the top of page 23.

3 Silicon and germanium are on the borderline between metals and non-metals. These elements are semiconductors, intermediate between metals, which are electrical conductors, and non-metals, which are non-conductors of electricity. Semiconductors are vital to the computer industry.

4 Group 7 is a set of very reactive non-metallic elements called the **halogens**. They react with metals to form salts; see the table on page 23.

5 When Mendeleev drew up his periodic table in 1871, only 55 elements were known. He left gaps in the table and predicted that new elements would be discovered which would fit the gaps. When the noble gases were discovered, one by one, their atomic numbers placed them in between Groups 1 and 7, and a new Group 0 had to be created for them.

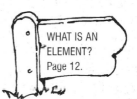

WHAT IS AN ELEMENT? Page 12.

3.2 Metallic and non-metallic elements

metallic elements	non-metallic elements
physical properties	*physical properties*
solids except for mercury	solids and gases, except for bromine (which is a liquid)
dense, hard	Most of the solid elements are softer than metals (diamond is exceptional).
A smooth metallic surface is shiny; many metals tarnish in air.	Most non-metallic elements are dull (diamond is exceptional).
The shape can be changed without breaking by the application of force – either compression, as in hammering, or tension, as in stretching, e.g. drawing out into a wire.	Many non-metallic elements are brittle – they break when a force is applied.
conduct heat (although highly polished surfaces reflect heat)	are poor thermal conductors
are good electrical conductors	are poor electrical conductors, except for graphite; some, e.g. silicon, are semiconductors
are sonorous – make a pleasing sound when struck	are not sonorous
The properties of metals derive from the metallic bond – see page 75.	
chemical properties	*chemical properties*
many displace hydrogen from dilute acids to form salts	do not react with acids (except for oxidising acids)
The metal is the cation (positive ion) in the salts, e.g. Na^+, Ca^{2+}; some metals also form oxoanions, e.g. ZnO_2^{2-}, AlO_3^-.	form anions (negative ions), e.g. S^{2-}, and oxoanions, e.g. SO_4^{2-}
form basic oxides and hydroxides, e.g. Na_2O, NaOH, CaO, $Ca(OH)_2$	form acidic oxides, e.g. CO_2, SO_2, or neutral oxides, e.g. CO, NO
The chlorides are ionic solids, e.g. $MgCl_2$, NaCl.	The chlorides are covalent gases or volatile liquids, e.g. HCl, CCl_4.

Characteristics of metallic and non-metallic elements

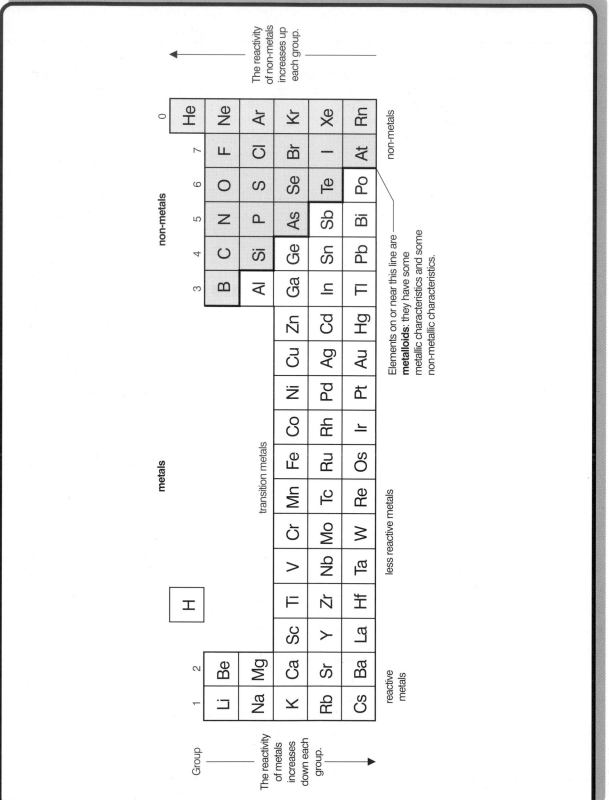

The periodic table

metal	reaction with air	reaction with water	reaction with dilute hydrochloric acid	trend	FACTS
Group 1 the alkali metals					
lithium sodium potassium rubidium caesium	Burn vigorously to form the strongly basic oxide M_2O which dissolves in water to give the strong alkali MOH.	React vigorously to form hydrogen and a solution of the strong alkali MOH.	The reaction is dangerously violent.	The vigour of all these reactions increases down the group.	
Group 2 the alkaline earths					
beryllium magnesium calcium strontium barium	Burn to form the strongly basic oxides MO, which are sparingly soluble or insoluble.	Reacts very slowly. Burns in steam. React readily to form hydrogen and the alkali $M(OH)_2$.	React readily to give hydrogen and a salt, e.g. MCl_2.	The vigour of all these reactions increases down the group. Group 2 elements are less reactive than Group 1.	
Transition metals					
iron zinc copper	When heated, form oxides without burning. The oxides and hydroxides are weaker bases than those of Groups 1 and 2 and are insoluble.	Iron rusts slowly. Iron and zinc react with steam to form hydrogen and the oxide. Copper does not react.	Iron and zinc react to give hydrogen and a salt. Copper does not react.	Transition metals are less reactive than Groups 1 and 2. In general, their compounds are coloured; they are used as catalysts.	

Some reactions of metals

Note
M stands for the symbol of a metallic element. Dilute sulphuric acid reacts with metals in the same way as dilute hydrochloric acid.

halogen	state at room temperature	reaction with sodium	reaction with iron	trend	FACTS
fluorine	gas	explosive	explosive		
chlorine	gas	Heated sodium burns in chlorine to form sodium chloride.	Reacts vigorously with hot iron to form iron(III) chloride.	The vigour of these reactions decreases down the group.	
bromine	liquid	Reacts less vigorously to form sodium bromide.	Reacts less vigorously to form iron(III) bromide.		
iodine	solid	Reacts less vigorously than bromine to form sodium iodide.	Reacts less vigorously than bromine to form iron(II) iodide.		

Some reactions of the halogens

3.3 Atoms

Elements are composed of atoms. Different elements have different types of atoms. The atoms of an element are different from the atoms of all other elements. Atoms are very small: the smallest atom is the hydrogen atom. There are 6×10^{23} atoms in 1 g of hydrogen. A uranium atom is 238 times as heavy as a hydrogen atom.

Atoms are composed of even smaller particles, subatomic particles. These are:

★ **protons**, positively charged, mass 1 atomic mass unit (1 amu)
★ **neutrons**, uncharged, mass 1 atomic mass unit (1 amu)
★ **electrons**, negatively charged, mass 0.0005 amu.

The atom is uncharged because the positive charge of the nucleus is equal to the sum of the negative charges of the electrons. The diagram below shows how protons, neutrons and electrons are arranged in the atom.

The electrons moving in orbits further away from the nucleus have more energy than those close to the nucleus. The orbits of lower energy are filled first. A group of orbits of similar energy is called a **shell**. The first shell can hold up to two electrons.

The second shell and the following outer shells can each hold up to eight electrons.

If you know the atomic number of an element, you can work out the arrangement of electrons. The lower energy levels are filled before the higher energy levels. The arrangements of electrons in an atom of carbon (atomic number 6) and an atom of magnesium (atomic number 12) are shown here.

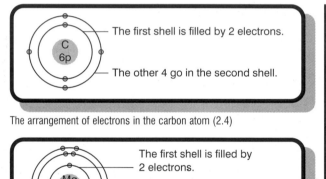

The first shell is filled by 2 electrons.

The other 4 go in the second shell.

The arrangement of electrons in the carbon atom (2.4)

The first shell is filled by 2 electrons.

The second shell is filled by 8 electrons.

The other 2 go in the third shell.

The arrangement of electrons in the magnesium atom (2.8.2)

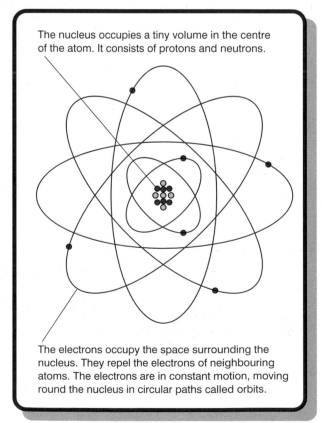

The nucleus occupies a tiny volume in the centre of the atom. It consists of protons and neutrons.

The electrons occupy the space surrounding the nucleus. They repel the electrons of neighbouring atoms. The electrons are in constant motion, moving round the nucleus in circular paths called orbits.

The arrangement of particles in the atom

Mass
Mass of proton
= mass of neutron
= 1 atomic mass unit, a.m.u.
Mass of electron
= 0.0005 a.m.u.

Some elements consist of **isotopes** –
forms of the element which have the
same number of protons but different
numbers of neutrons (the same
atomic number but different mass
numbers.)

Charge
Negative charge
on electron = –1 elementary
charge unit.
Positive charge
on proton = +1.
Neutrons are uncharged.

THE ATOM consists of
subatomic particles:
protons, **neutrons**
and **electrons**.

The protons and neutrons
are located in the **nucleus**
of the atom. The electrons
are present in the space
outside the nucleus
(see diagram opposite).

Chlorine consists of two kinds of
atoms, one with 17 protons and 18
neutrons, called chlorine-35, and one
with 17 protons and 20 neutrons,
called chlorine-37. The **isotopes** can
be written as

mass number $\rightarrow$ 35
 Cl $\leftarrow$ symbol
atomic number $\rightarrow$ 17
and $^{37}_{17}$ Cl.

Atomic number of element
= number of protons in the nucleus of
an atom of the element
= number of electrons in the atom.
Mass number of atom = sum of
number of protons + number of neutrons.
Number of neutrons = mass no.
– atomic no.

Relative atomic mass (RAM) =
mass of one atom of element ÷ $\frac{1}{12}$mass
of one atom of carbon-12.
Relative formula mass (RFM) =
mass of one molecule or one formula
unit of a compound ÷ $\frac{1}{12}$mass of one
atom of carbon-12.

Electrons occupy orbitals. They are
arranged in shells of orbitals, with
orbitals of lowest energy filled first
(see overleaf). The **electron arrangement**
of an atom can be predicted from its
atomic number, e.g. Ca with $Z = 20$
has the arrangement 2.8.8.2.

The isotopes of carbon are $^{12}_{6}$C, $^{13}_{6}$C and $^{14}_{6}$C.

The isotopes of hydrogen are $^{1}_{1}$H, $^{2}_{1}$H
and $^{3}_{1}$H.

Concept map: the nature of the atom

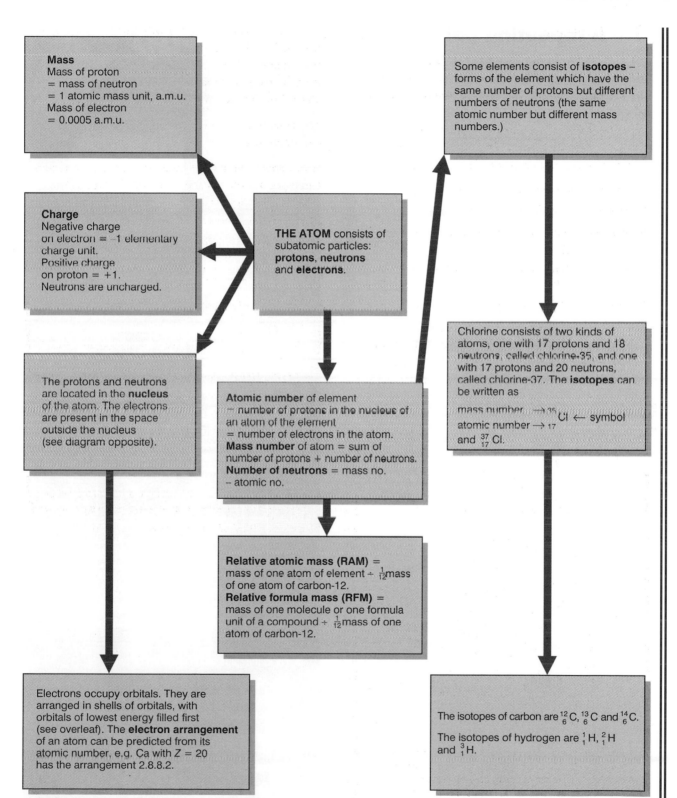

3.4 A repeating pattern

We have covered the structure of the atom, the atomic numbers of the elements and the arrangement of electrons in the elements. How does all this fit in with the periodic table?

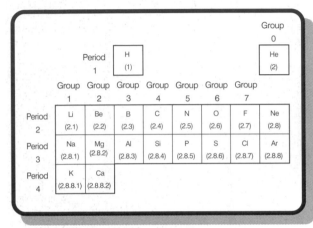

A section of the periodic table

You can see that the arrangement above has the following features:

★ The elements are listed in order of increasing atomic number.

★ Elements which have the same number of electrons in the outermost shell fall into the same **group** (vertical column) of the periodic table.

★ The noble gases are in Group 0. For the rest of the elements, the group number is the number of electrons in the outermost shell.

★ The first **period** (horizontal row) contains only hydrogen and helium. The second period contains the elements lithium to neon. The third period contains the elements sodium to argon.

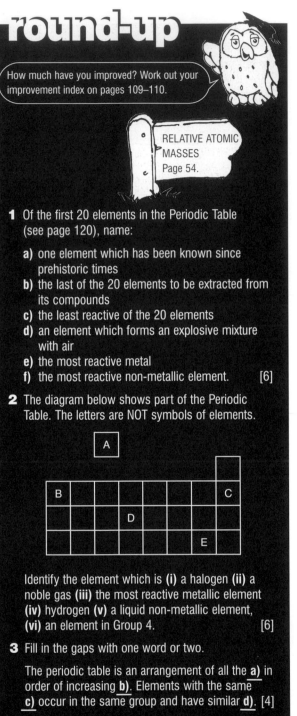

round-up

How much have you improved? Work out your improvement index on pages 109–110.

RELATIVE ATOMIC MASSES Page 54.

1 Of the first 20 elements in the Periodic Table (see page 120), name:

a) one element which has been known since prehistoric times

b) the last of the 20 elements to be extracted from its compounds

c) the least reactive of the 20 elements

d) an element which forms an explosive mixture with air

e) the most reactive metal

f) the most reactive non-metallic element.   [6]

2 The diagram below shows part of the Periodic Table. The letters are NOT symbols of elements.

Identify the element which is **(i)** a halogen **(ii)** a noble gas **(iii)** the most reactive metallic element **(iv)** hydrogen **(v)** a liquid non-metallic element, **(vi)** an element in Group 4.   [6]

3 Fill in the gaps with one word or two.

The periodic table is an arrangement of all the **a)** in order of increasing **b)**. Elements with the same **c)** occur in the same group and have similar **d)**. [4]

4 a) Explain why an atom is unchanged although it contains particles which are electrically charged.   [1]

b) Where in the atom **(i)** the positvely charged particles **(ii)** the negatively charged particles?   [2]

5 Mendeleev predicted that the properties of germanium long before it was discovered. How was he able to do this? [2]

6 Refer to the Periodic Table on page 120. Complete the two tables below.

element	symbol	atomic number	electron arrangement
Helium			
Boron			
Aluminium			
Carbon			
Nitrogen			
Fluorine			

[18]

atom or ion	No. of protons	No. of neutrons	No. of electrons
a) $^{7}_{3}Li$			
b) $^{7}_{3}Li^{+}$			
c) $^{24}_{12}Mg^{2+}$			
d) $^{19}_{9}F$			
e) $^{32}_{16}S^{2-}$			

[15]

7 Identify the quantity or quantities **(i)** that are equal to the atomic number of an element **(ii)** that differ between isotopes of an element **(iii)** that decide the chemical properties of an element:

A the number of protons in an atom
B the number of neutrons in an atom
C the number of electrons in an atom
D the number of electrons in the outer shell of an atom. [4]

8 a) The element Q forms ions Q^{2+}. Is Q metallic or non-metallic? [1]
b) The element R forms an acidic oxide RO_2. Is R metallic or non-metallic? [1]
c) The element E forms a crystalline chloride ECl_2. Is E metallic or non-metallic? [1]
d) The element G forms a stable hydride HG. Is G metallic or non-metallic? [1]

9 Contrast four physical properties for sulphur and copper (a non-metallic element and a metallic element). [4]

10 a) How many times heavier is one atom of aluminium than one atom of hydrogen? [1]
b) How many times heavier is one atom of mercury than one atom of calcium? [1]
c) What is the ratio:
$$\frac{\text{mass of one Fe atom}}{\text{mass of one Br atom?}}$$ [1]
d) How many atoms of nitrogen equal the mass of one atom of bromine? [1]

11 Write the symbol, with mass number and atomic number, for each of the following isotopes:
a) arsenic (atomic number 33 and mass number 75) [1]
b) uranium-235, uranium-238 and uranium-239 (atomic number 92). [3]

12 The electron arrangement of phosphorus is (2.8.5). Sketch the arrangement of electrons in the atom, as in the diagrams above. [1]

13 Sketch the arrangements of electrons in the atoms of
a) B (atomic number 5)   **b)** N (atomic number 7)
c) F (atomic number 9)   **d)** Al (atomic number 13). [4]

Well done if you've improved. Don't worry if you haven't. Take a break and try again.

How atoms combine

preview

At the end of this topic you will understand:

- how atoms of non-metallic elements combine
- what these atoms gain by sharing a pair of electrons
- why this shared pair bonds the atoms as a single covalent bond.

How much do you already know? Work out your score on page 110.

Test yourself

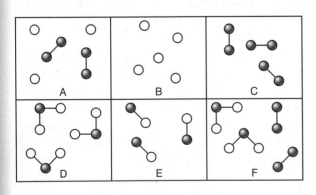

1 Which diagram represents **(i)** an element composed of diatomic molecules **(ii)** a mixture of a compound and an element? [2]

2 Name **(i)** the compound that contains two elements only **(ii)** the convalent compound.
   A ammonium phosphate
   B sodium sulphide
   C sodium sulphate
   D aluminium nitrate
   E copper carbonate
   F sulphur chloride [2]

3 Which compounds are **(i)** covalent **(ii)** dissolve in water to give an acidic solution **(iii)** dissolve in water to give an alkaline solution?
   A CuO
   B Al_2O_3
   C K_2O
   D SO_2
   E SiO_2
   F NO [5]

4.1 The chemical bond

Elements and compounds consist of atoms joined together: combined. The force holding two atoms together is a **chemical bond**. There are two chief types of chemical bond:

- the **ionic bond**, formed when metals and non-metallic elements combine (see Topic 7)
- the **covalent bond**, formed when atoms of non-metallic elements combine.

4.2 The covalent bond

Two atoms form a covalent bond between them by sharing a pair of electrons. Elemental gases, except for the noble gases, consist of diatomic molecules, e.g.

- two hydrogen atoms bond to form a molecule, H_2
- two chlorine atoms bond to form a molecule, Cl_2

These are **single covalent bonds**. Chlorine has a **valency** of one: a chlorine atom can form only a single covalent bond.

Some atoms share two pairs of electrons to form a **double covalent bond**. And some others share three pairs of electrons to form a **triple bond**, e.g.

- the oxygen molecule, O_2
- the nitrogen molecule, N_2

Atoms of *different* elements can share pairs of electrons to form covalent bonds. The bonded groups of atoms are called **molecules**.

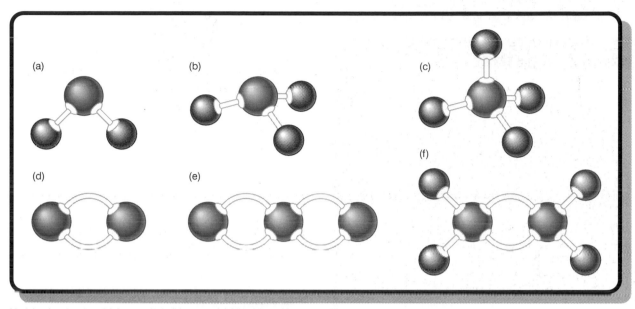

Models of molecules of **(a)** water, H_2O **(b)** ammonia, NH_3 **(c)** methane, CH_4 **(d)** oxygen, O_2 **(e)** carbon dioxide, CO_2 **(f)** ethene, C_2H_4

Note the number of bonds that each atom forms. Hydrogen and chlorine have valency 1, oxygen has valency 2, nitrogen has valency 3, carbon has valency 4.

Some elements can use more than one valency. Carbon has a valency of 2 in carbon monoxide, CO, and a valency of 4 in carbon dioxide, CO_2. Sulphur has a valency of 4 in sulphur dioxide, SO_2, and a valency of 6 in sulphur trioxide, SO_3.

4.3 Formulae

Instead of drawing molecules (as above) we write formulae for compounds. The formula consists of **symbols** which show the elements that are present in the compound and **numbers,** which show the ratio of atoms of the different elements. The formula H_2SO_4 for sulphuric acid shows that the elements present are hydrogen, sulphur and oxygen and that the ratio of atoms is 2 H atoms: 1 S atom: 4 O atoms.

4.4 What do atoms gain by sharing electrons?

The formation of covalent bonds happens only with non-metallic elements. A chlorine atom has 7 electrons in its outer shell. One more electron would 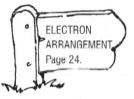 give it a full shell of 8 electrons. We know this is a very stable (unreactive) arrangement because the noble gas argon has this arrangement and argon takes part in no chemical reactions. When two chlorine atoms share a pair of electrons the outer shells overlap and both chlorine atoms have 8 electrons in their outer shells.

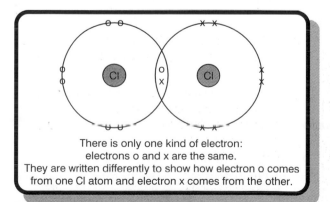

There is only one kind of electron: electrons o and x are the same.
They are written differently to show how electron o comes from one Cl atom and electron x comes from the other.

4.5 Why does sharing a pair of electrons bond two atoms together?

The two chlorine nuclei are positively charged. The shared pair of electrons is negatively charged. The shared pair is attracted to both nuclei and so holds them together.

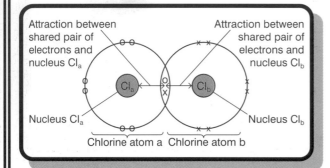

I always knew that nuts came in shells. Now I know that electrons come in shells too, and 8 electrons make a full shell, and every atom wants to get a full shell! Why? Because with a full shell the atom is stable.

4.6 Covalent bonding

Example 1

Hydrogen + fluorine → hydrogen fluoride, HF

The hydrogen atom shares its electron with the fluorine atom. H has a full shell of two electrons, the same arrangement as helium.

The fluorine atom shares one of its electrons with the hydrogen atom. F has a full shell of eight electrons, the same arrangement as neon.

The formation of hydrogen fluoride

The shared pair of electrons is attracted to the hydrogen nucleus and to the fluorine nucleus, and bonds the two nuclei together.

Example 2

Hydrogen + oxygen → water, H_2O

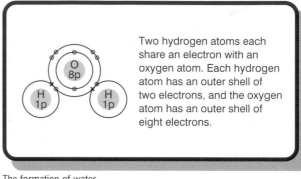

Two hydrogen atoms each share an electron with an oxygen atom. Each hydrogen atom has an outer shell of two electrons, and the oxygen atom has an outer shell of eight electrons.

The formation of water

Example 3

Nitrogen + hydrogen → ammonia, NH_3

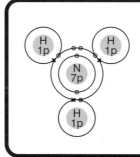

One nitrogen atom shares three electrons, one with each of three hydrogen atoms. The nitrogen atom has eight electrons in its outer shell (like neon), and each hydrogen has two electrons in its outer shell (like helium).

The formation of ammonia

Example 4

Carbon + hydrogen → methane, CH_4

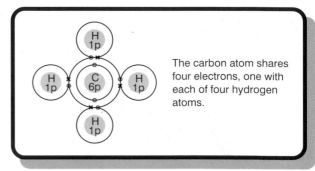

The carbon atom shares four electrons, one with each of four hydrogen atoms.

The formation of methane

Example 5

Carbon + oxygen → carbon dioxide, CO_2

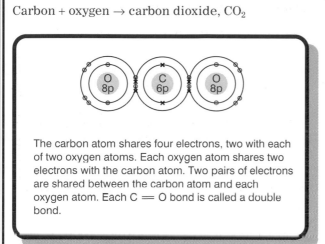

The carbon atom shares four electrons, two with each of two oxygen atoms. Each oxygen atom shares two electrons with the carbon atom. Two pairs of electrons are shared between the carbon atom and each oxygen atom. Each C $=$ O bond is called a double bond.

The formation of carbon dioxide

4.7 The shapes of molecules

Since covalent bonds are shared pairs of electrons, they are regions of negative charge. The charges repel one another and take up positions in space so that they are as far apart as possible. The way in which the covalent bonds are directed in space decides the shape of the molecule.

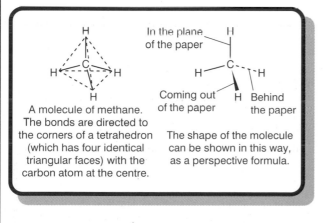

A molecule of methane. The bonds are directed to the corners of a tetrahedron (which has four identical triangular faces) with the carbon atom at the centre.

The shape of the molecule can be shown in this way, as a perspective formula.

Pairs of electrons which are not involved in bond formation also help to decide the shape of the molecule. The nitrogen atom has 5 electrons in the outer shell. In NH_3, N shares three electrons with H atoms and has an unshared pair of electrons. The three covalent bonds and the

unshared pair of electrons are directed towards the corners of a tetrahedron. The oxygen atom has 6 electrons in the outer shell. In H_2O, forms two covalent bonds and has two unshared pairs of electrons. The two covalent bonds and the two unshared pairs of electrons are also directed towards the corners of a tetrahedron.

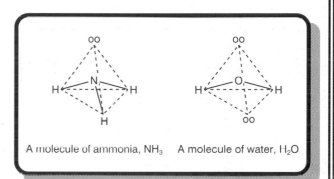

A molecule of ammonia, NH_3 A molecule of water, H_2O

round-up

How much have you improved?
Work out your improvement index on page 110.

1 Draw diagrams to show the shapes of these molecules:

   H_2O, HCl, CO_2, NCl_3, CCl_4, $SiCl_4$, H_2S, F_2, CF_4
   [10]

2 Draw diagrams to show how the outer electrons are shared to form covalent bonds in the molecules of HBr, F_2 and H_2S.

   Electron arrangements are H (1), Br (2, 8, 18, 7), F (2.7), S (2.8.6). [3]

3 Explain why sharing a pair of electrons holds two atoms together. [2]

4 Why do two atoms of argon not combine to form a molecule Ar_2? [1]

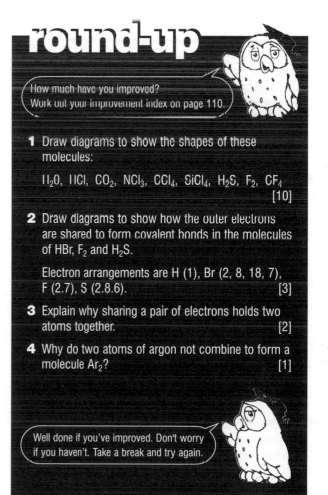

Well done if you've improved. Don't worry if you haven't. Take a break and try again.

Fuels

preview

At the end of this topic you will know about:

- fossil fuels: coal, oil and natural gas
- exothermic and endothermic reactions
- the fuel crisis
- pollution associated with fossil fuels.

How much do you already know? Work out your score on page 110.

Test yourself

1 Why are coal and oil called 'fossil fuels'? [3]

2 What is most of the world's coal used for? [1]

3 Crude oil can be separated into useful fuels and other substances.
   a) Name the process which is used. [1]
   b) Name four fuels obtained from crude oil. [4]
   c) Name two other useful substances separated from crude oil. [2]

4 Describe tests for:
   a) oxygen [2]
   b) carbon dioxide [2]
   c) water. [2]

5 Explain what is meant by 'cracking'. [4]

6 Divide the following list of reactions into
   a) exothermic reactions b) endothermic reactions: [4]
   photosynthesis, combustion, cracking of hydrocarbons, respiration.

5.1 Fossil fuels

Coal

Coal is one of the fuels we describe as **fossil fuels**. It was formed from dead plant material decaying slowly over millions of years under the pressure of deposits of mud and sand. Coal is a complicated mixture of carbon, hydrocarbons and other compounds. Much of the coal used in the world is burned in power stations. The main combustion products are carbon dioxide and water.

COMBUSTION Page 34.

Petroleum oil and natural gas

Petroleum oil (usually called simply oil) and natural gas are fossil fuels: they are the remains of sea animals which lived millions of years ago. Decaying slowly under the pressure of layers of mud and silt, the organic part of the creatures' bodies turned into a mixture of hydrocarbons: petroleum oil. The sediment on top of the decaying matter became compressed to form rock, so oil is held in porous oil-bearing rock. Natural gas is always formed in the same deposits as oil.

The economic importance of oil

Industrialised countries depend on fossil fuels for transport, for power stations and for manufacturing industries. Scotland benfits from the exploitation of North Sea oil and gas. The petrochemicals industry makes a vast number of important chemicals from oil including fertilisers, herbicides, insecticides and the raw materials needed by the pharmaceutical industry. Earth's deposits of coal, oil and gas are **finite resources**. When we have used them there will be no more forthcoming. The economies of all industrial countries will depend on alternative energy sources.

5.2 Fuels from petroleum

Fractional distillation

Crude oil is separated by fractional distillation into a number of important fuels. Each fraction is collected over a certain boiling point range. Each fraction is a mixture of hydrocarbons (compounds which consist of hydrogen and carbon only). The use that is made of each fraction depends on these factors, all of which increase with the size of the molecules:

- its **boiling point range**: the higher the boiling point range, the more difficult it is to vaporise in a vehicle engine.
- its **viscosity**: the more viscous a fraction is, the less easily it flows.
- its **ignition temperature**: the less easily a fraction ignites, the less flammable it is.

Cracking

We use more naphtha, petrol and kerosene than heavy fuel oil. The technique called **cracking** is used to convert the high boiling point range fractions into the lower boiling point range fractions petrol and kerosene and unsaturated hydrocarbons.

$$
\text{vapour of hydrocarbon with large molecules and high boiling point} \xrightarrow[\substack{\text{passed over a} \\ \text{heated catalyst} \\ \text{e.g. } Al_2O_3}]{\text{cracking}} \text{mixture of hydrocarbons with smaller molecules and low boiling point, and hydrogen. The mixture is separated by fractional distillation.}
$$

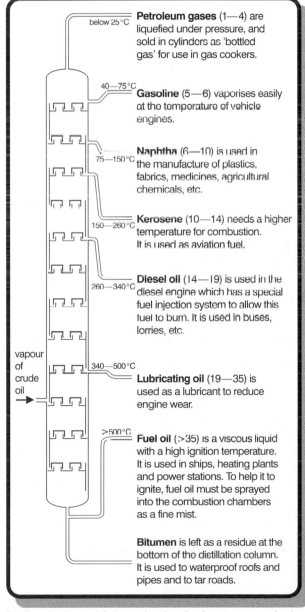

Petroleum gases (1—4) are liquefied under pressure, and sold in cylinders as 'bottled gas' for use in gas cookers. *below 25 °C*

Gasoline (5—6) vaporises easily at the temperature of vehicle engines. *40—75 °C*

Naphtha (6—10) is used in the manufacture of plastics, fabrics, medicines, agricultural chemicals, etc. *75—150 °C*

Kerosene (10—14) needs a higher temperature for combustion. It is used as aviation fuel. *150—260 °C*

Diesel oil (14—19) is used in the diesel engine which has a special fuel injection system to allow this fuel to burn. It is used in buses, lorries, etc. *260—340 °C*

Lubricating oil (19—35) is used as a lubricant to reduce engine wear. *340—500 °C*

Fuel oil (>35) is a viscous liquid with a high ignition temperature. It is used in ships, heating plants and power stations. To help it to ignite, fuel oil must be sprayed into the combustion chambers as a fine mist. *>500 °C*

Bitumen is left as a residue at the bottom of the distillation column. It is used to waterproof roofs and pipes and to tar roads.

vapour of crude oil →

Petroleum fractions and their uses (number of carbon atoms per molecule is shown in brackets)

5.3 Air

Fuels need oxygen or air to burn. The composition of air is shown in the diagram below.

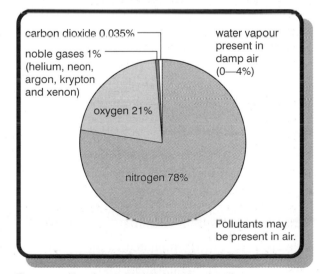

carbon dioxide 0.035%

noble gases 1% (helium, neon, argon, krypton and xenon)

water vapour present in damp air (0—4%)

oxygen 21%

nitrogen 78%

Pollutants may be present in air.

The composition of pure, dry air in percentage by volume

All fuels burn more rapidly in pure oxygen than in air. The same quantity of heat is released, but it is released in a shorter time. This is the basis of a **test for oxygen**: oxygen relights a glowing splint.

5.4 Products of combustion

When hydrocarbons burn in sufficient oxygen the products are carbon dioxide and water. The diagram below shows an experiment to illustrate this. If there is insufficient oxygen, incomplete combustion takes place and carbon monoxide and carbon are formed in addition.

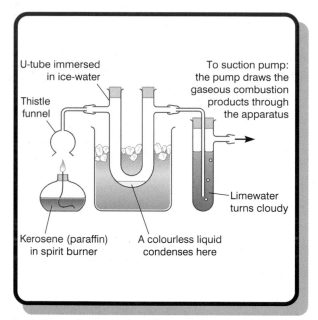

U-tube immersed in ice-water

Thistle funnel

To suction pump: the pump draws the gaseous combustion products through the apparatus

Limewater turns cloudy

Kerosene (paraffin) in spirit burner

A colourless liquid condenses here

Testing the combustion products of a hydrocarbon fuel, e.g. kerosene or candle wax (NOT petrol)

5.5 Energy and chemical reactions

Exothermic reactions

1 **Combustion**: the combustion of hydrocarbons is an exothermic reaction – heat is given out.

methane + oxygen → carbon dioxide + water;
heat is given out

octane + oxygen → carbon dioxide + water;
heat is given out

★ An oxidation reaction in which heat is given out is **combustion**.

★ Combustion accompanied by a flame is **burning**.

★ A substance which is oxidised with the release of energy is a **fuel**.

2 **Respiration**: our bodies obtain energy from the oxidation of foods, e.g. glucose, in cells. This process is called **cellular respiration**.

glucose + oxygen → carbon dioxide + water;
energy is given out

Endothermic reactions

1 **Photosynthesis**: plants convert carbon dioxide and water into sugars in the process of photosynthesis.

catalysed by chlorophyll
carbon dioxide + water → glucose + oxygen;
energy of sunlight is taken in

2 **Thermal decomposition**; for example the cracking of hydrocarbons and the decomposition of calcium carbonate:

$$\text{calcium carbonate} \xrightarrow{\text{heat}} \text{calcium oxide} + \text{carbon dioxide};$$
heat is taken in

Tests

Carbon dioxide turns limewater (calcium hydroxide solution) cloudy.

Carbon dioxide + calcium hydroxide → calcium carbonate + water

$CO_2(g) + Ca(OH)_2(aq) \rightarrow CaCO_3(s) + H_2O(l)$

Water turns anhydrous copper(II) sulphate blue.

anhydrous copper(II) sulphate (white) + water → copper(ll) sulphate crystals (blue)

$CuSO_4(s) + 5H_2O(l) \rightarrow CuSO_4, 5H_2O(s)$

Pure water has a boiling point 100 °C at 1 atmosphere and freezing point 0 °C at 1 atmosphere.

5.6 The fuel crisis

The present way of life in industrialised countries depends largely on oil and gas. Agriculture depends on these too: the petrochemicals industry provides farmers with fertilisers to increase their yields and pesticides to protect their crops from insects and weeds. The manufacturing industries, which supply most of our possessions, depend on oil. All the cars, trains, boats and planes which transport us would be useless without oil. Without the petrochemicals industry, we would have few modern drugs and medicines. Without oil and gas, the whole world economy would collapse.

The problem is that we are consuming energy faster and faster. The Earth's reserves of coal, oil and natural gas are limited. These fossil fuels were made millons of years ago. Once we have used up the Earth's deposits, there will be no more. Scientists estimate that reserves of oil will run out in 30 to 40 years from now. This problem is what people call the **fuel crisis** or the **energy crisis**. The solution to the problem is to find new sources of energy.

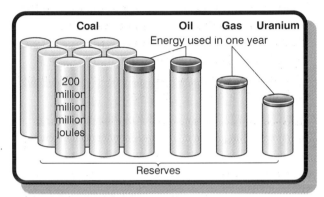

Estimated reserves of fossil fuels in the world

	coal	oil	natural gas
world reserves	200–250 years	30–40 years	20 years
Effect on the environment	Coal mines create ugly spoil heaps, which can be dangerous. Open cast mining is a blight on the landscape. Mining is a dangerous job.	The sea is polluted when tankers have accidents and when ships illegally wash out their tanks at sea.	
Combustion	Pollutants discharged into the air are: sulphur dioxide, carbon monoxide, oxides of nitrogen, hydrocarbons and soot.	Pollutants discharged into the air are: sulphur dioxide, carbon monoxide, oxides of nitrogen, hydrocarbons and soot.	Natural gas is a clean-burning fuel.

A comparison of three fossil fuels

Some alternative sources of energy

New fuels from coal

One way of tackling the problem of diminishing resources of oil is to make more use of coal.

Already there are methods of making **a)** liquid fuels and **b)** gaseous fuels from coal.

a)

$$\text{coal (carbon + hydrocarbons)} + \text{hydrogen} \xrightarrow[\text{catalyst}]{\text{high temperature and pressure}} \text{liquid hydrocarbon fuel}$$

The fuel produced can be used to power diesel engines.

b)

(1) Heat

coal → fuel gas (carbon monoxide + hydrogen + methane)

(2) React with steam

Nuclear Power

One solution to the energy crisis is to obtain all the world's power needs from nuclear reactors. This solution poses other problems. There is the danger of accidents in nuclear reactors and the need to find safe ways of storing long-lived radioactive waste material. Uranium, used as fuel in reactors, is a non-renewable fuel. When all the Earth's resources of uranium have been used up, there will be no more for us to use.

Renewable energy sources

There are some sources of energy which will never run out. These are the **renewable energy sources**. They derive their energy from the Sun. We cannot use them up because the Sun constantly renews them.

★ **Solar energy** – this is the energy derived from sunlight. Solar energy can be converted into electricity by certain crystals in photoelectric cells.

★ **Wind power** – aerogenerators are driven by the wind. They make a lot of noise so they cannot be placed near housing. The electricity generated has to be carried overland in power cables.

★ **Hydroelectricity** – hydro-electric power is obtained by using water from waterfalls and fast-flowing rivers to drive turbines.

★ **Tidal power** – the energy of the tides can also be converted into electrical energy. The rising tide must be trapped behind a 'barrage'. Later the high tide is released to the sea through turbines.

★ **Wave power** – ocean waves are another source of power.

★ **Geothermal energy** – deep down below the surface of the Earth, the temperature is much higher than at the surface. In some places however, the Earth's crust is thin, and the hot rocks are only hundreds of metres down. In such places water can be pumped down over the hot rock (at about 200 °C), where it turns into steam. The steam returns to the surface to be used to drive turbines.

★ **Biomass** – material from living organisms is called biomass. It can be converted into fuels. Refuse from homes and factories could provide enough biogas to make a substantial contribution, say 10%, to our energy needs.

fossil fuels	renewable energy sources
The Earth's reserves are limited.	The supply will never run out because the energy comes from the Sun.
The burning of fossil fuels causes many types of pollution.	Harnessing renewable energy sources causes no pollution.
Coal mines and oil well are less expensive to build than plants for utilising renewable energy.	Large investment is needed to build solar power stations, geothermal power stations etc. Often a large area of land is needed, e.g. for wind machines.

Comparison of fossil fuels and renewable energy sources

5.7 The problem of air pollution

Carbon monoxide

Source: most of the carbon monoxide in the air comes from vehicle engines, where it is formed by the incomplete combustion of petrol. Soil organisms remove carbon monoxide from the air. However, in cities, where the concentration of carbon monoxide is high, there is little soil to remove it.

Effects: carbon monoxide combines with haemoglobin, the red pigment in the blood, and prevents haemoglobin from combining with oxygen. At a level of 1%, carbon monoxide will kill quickly; at lower levels, it causes headaches and dizziness and affects reaction times. Being colourless and odourless, carbon monoxide gives no warning of its presence.

Solutions to the problem may come from

- tuning vehicle engines to use more air and produce only carbon dioxide and water. However, this increases the emission of oxides of nitrogen.
- fitting vehicles with catalytic converters
- using fuels which burn more cleanly than hydrocarbons, e.g. ethanol.

Sulphur dioxide

Sources: major sources of the sulphur dioxide in the air are

- the extraction of metals from sulphide ores
- the burning of coal, which contains 0.5–5% sulphur, mostly in electricity power stations
- oil-burning power stations, because fuel oil contains sulphur compounds.

ACID RAIN
Page 53.

Effects: sulphur dioxide is a colourless gas with a very penetrating and irritating smell. Atmospheric sulphur dioxide is thought to contribute to bronchitis and lung diseases. It is a cause of acid rain.

Smoke, dust and grit

Particles enter the air from natural sources such as dust storms, forest fires and volcanic eruptions. Coal-burning power stations, incinerators, industries and vehicles add to the pollution. When smoke particles mix with fog, **smog** is formed. Smog contains sulphuric acid, which has been formed from sulphur dioxide in the smoke. Breathing smog makes the lungs produce mucus, making it more difficult to breathe.

Methods of removing particles include:

- using sprays of water to wash out particles from waste gases
- passing exhaust gases through filters
- using electrostatic precipitators which attract particles to charged plates.

The exhaust gases of vehicles are not treated by any of these methods.

Oxides of nitrogen

Source: when fuels are burned in air, the temperature rises. Some of the nitrogen and oxygen in the air combine to form nitrogen monoxide, NO, and nitrogen dioxide, NO_2. This mixture (shown as NO_x) is emitted by power stations, factories and vehicles.

Effects: nitrogen monoxide is soon converted into nitrogen dioxide which is highly toxic, and which contributes to the formation of acid rain.

Solution to the problem: the presence of a catalyst (platinum) brings about the reaction:

$$\text{nitrogen monoxide} + \text{carbon monoxide} \rightarrow \text{nitrogen} + \text{carbon dioxide}$$

$$2NO(g) + 2CO(g) \rightarrow N_2(g) + 2CO_2(g)$$

The **catalytic converters** which are now fitted in the exhausts of cars reduce the emission of oxides of nitrogen in this way. Unleaded petrol must be used because lead compounds in the exhaust gases would stop the catalyst working.

Hydrocarbons

Sources: the hydrocarbons in the air come from natural sources, such as the decay of plant material (85%), and from vehicles (15%).

Effects: in sunlight, hydrocarbons react with oxygen and oxides of nitrogen to form **photochemical smog**. This contains irritating and toxic compounds.

Solutions to the problem: if the air supply in a vehicle engine is increased, the petrol burns completely. However, at the same time, the formation of NO_X increases. A solution may be found by running the engine at a lower temperature and employing a catalyst to promote combustion.

Lead

Sources: lead compounds enter the air from the combustion of coal, the roasting of metal ores and from vehicle engines. Since the introduction of unleaded petrol, the level of lead in the atmosphere has fallen.

Effects: lead causes depression, tiredness, irritability and headaches. Higher levels of lead cause damage to the brain, liver and kidneys.

5.8 Pollution of water

Pollution by oil

Modern oil tankers are huge, each carrying up to 500 000 tonnes of crude oil.

If a tanker has an accident at sea, oil is spilt, and a huge oil slick floats on the surface of the ocean. It is very slowly oxidised by air and decomposed by bacteria. While the oil slick remains, it poisons fish and glues the feathers of sea birds together so that they cannot fly. When the oil slick washes ashore, it fouls beaches. The following methods can be used to deal with oil slicks:

1 **dispersants** – powerful detergents

2 **sinking** the oil by spreading it with e.g. powdered chalk

3 **absorption** in e.g. straw and polystyrene

4 **booms** placed in the water to prevent oil from spreading

5 **natural processes** if the oil spill occurs far out to sea.

5.9 The greenhouse effect

The Earth receives radiation from the Sun, and also radiates heat into space. Carbon dioxide and water vapour reduce the escape of heat energy from the Earth by means of the **greenhouse effect** (see diagram below). Without these 'blankets' of water vapour and carbon dioxide, the temperature of the Earth's surface would be at −20°C, and life on Earth would be impossible.

The combustion of fossil fuels is causing an increase in the level of carbon dioxide at a rate which could raise the average temperature of the Earth. One result would be that the massive icecaps of the Arctic and Antarctic regions would slowly begin to melt. The levels of oceans would rise, and coastal areas would be flooded. A rise in temperature could decrease food production over vast areas, e.g. the mid-west USA and Russia.

Secondary effects would make matters worse. The increase in temperature would make more water vaporise from the oceans and drive out some of the carbon dioxide dissolved in the oceans to add to a still thicker greenhouse blanket. Other 'greenhouse gases' are methane, chlorofluorocarbons (CFCs), nitrogen oxides and ozone.

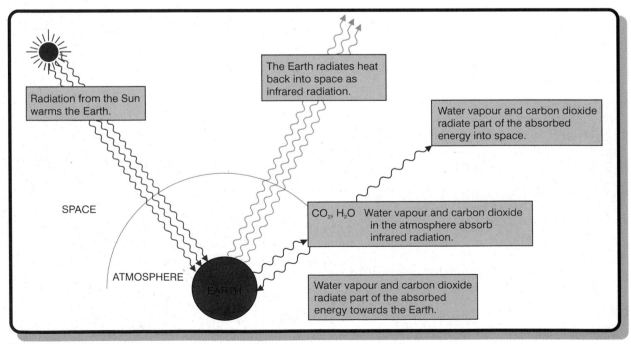

The greenhouse effect

round-up

How much have you improved?
Work out your improvement index on page 110–111.

1 What is meant by each of the terms underlined?
   a) fossil fuel [1]
   b) finite resource [1]
   c) renewable resource [1]
   d) fuel crisis [1]

2 a) Briefly explain how crude oil is separated into natural gas, gasoline, naphtha, kerosene, diesel oil and fuel oil. [2]
   b) State one use for each of the fractions. [6]
   c) Give three differences in properties between gasoline and fuel oil. [3]
   d) A barrel of oil yields less gasoline than fuel oil. The demand for gasoline is, however, greater than the demand for fuel oil. Explain why fuel oil cannot be used in car engines. [2]
   e) Briefly explain why an increase in the price of oil has a great impact worldwide. [4]
   f) Explain why the boiling points of the fractions increase steadily from gasoline to fuel oil. [2]
   g) What can you say about the change in **(i)** viscosity **(ii)** flammability from gasoline to fuel oil? [4]

3 Explain the danger to health from the presence in the air of **a)** carbon monoxide **b)** sulphur dioxide and **c)** particles of smoke and dust.
   Say how nature removes each of these pollutants from the air. [6]

4 a) How do oxides of nitrogen get into the air?
   b) What damage do they cause?
   c) What is the solution to this problem? [3]

5 a) Explain how the 'greenhouse effect' makes life on Earth possible. [2]
   b) Explain why people are worried about an 'enhanced greenhouse effect'. [2]
   c) If Earth warms up, what do people predict will happen
      (i) at the North Pole **(ii)** in Thailand
      (iii) in the mid-west of the USA? [3]

6 Explain why power stations have tall chimneys. Would the problem of pollution from power stations be solved by still taller chimneys? [2]

7 a) Why do cold countries, e.g. Sweden, suffer badly from acid rain? [1]
   b) In Sweden, the base calcium hydroxide is added to acid lakes. Write a word equation and a chemical equation for the reaction between calcium hydroxide and sulphuric acid in the lake water. [4]
   c) What name is given to this type of reaction? [1]

8 What chemical reactions take place between acid rain and **a)** iron railings **b)** marble statues **c)** fresh mortar? [2, 3, 2]

9 a) In petrol engines and diesel engines, hydrocarbons burn to form a number of products. What are these products? [4]
   b) What other substances are present in vehicle exhaust gases? [3]

10 Catalysts employed in vehicle exhausts include transition metals. Exhaust gases include CO and NO.
   a) What is a transition metal? Give an example of one that is used in this way. [2]
   b) Write an equation for the catalysed reaction between CO and NO. [4]

11 In a coal-fired power station sulphur dioxide is produced.
   a) What is the origin of the sulphur dioxide? [1]
   b) Why is it necessary to remove sulphur dioxide from the exhaust gases? [2]
   c) A method of removing sulphur dioxide is to pass the exhaust fumes from the power station through powdered limestone and oxidise the calcium sulphite produced to calcium sulphate. What mass of $CaCO_3$ is needed to convert the daily output of a power station of 560 tonnes SO_2 into $CaSO_4$? [2]

12 What are the advantages and disadvantages as a source energy of **(i)** coal and **(ii)** natural gas?

13 a) Which is the more reliable source of power: wind power or tidal power? [1]
   b) Clusters of aerogenerators must be sited away from residential areas. Why? [1]
   c) What effect does the building of a tidal power station have on a coastal region? [1]
   d) What effect does the building of a wave power barrage have on the business of an estuary? [1]

Hydrocarbons

preview

At the end of this topic you will:

- **know the general formulae and reactions of alkanes, cycloalkanes and alkenes.**

How much do you already know? Work out your score on page 111.

Test yourself

1 Draw the functional group of an alkene. Say in what type of reactions this functional group takes part. [2]

2 Unlike alkanes, alkenes are not used as fuels. Why is this? [2]

3 Why does ethene decolorise a solution of bromine? Give the formula of the product. [1]

4 a) Give the meanings of these terms: <u>hydrocarbon</u>, <u>saturated</u> hydrocarbon, <u>unsaturated</u> hydrocarbon. [3]
b) What is meant by the <u>cracking</u> of alkanes? [4]
c) State three reasons why it is an important industrial process. [4]
d) Why is a catalyst used? [2]

5 a) What is hydrogenation? [1]
b) What is the industrial importance of hydrogenation? [1]

6 a) Give the molecular formulae and structural formulae of propane and propene. [4]
b) Given samples of these two gases, how could you test to see which was which? [2]

• •

6.1 Alkanes

Most of the hydrocarbons in crude oil belong to the **homologous series** called **alkanes**; they are shown in the table (below). A homologous series is a set of compounds with similar chemical properties in which one member of the series differs from the next by a $-CH_2-$ group. Physical properties vary gradually as the size of the molecules increases. The boiling point increases with the number of carbon atoms per molecule. The alkane molecules are long chains of $-CH_2-$ groups. The chains line up side by side with attractive forces between them. The larger the molecules, the greater the forces of attraction between them. More energy has to be supplied to separate the molecules; that is the boiling point rises.

name	molecular formula	structural formula
methane	CH_4	CH_4
ethane	C_2H_6	CH_3CH_3
propane	C_3H_8	$CH_3CH_2CH_3$
butane	C_4H_{10}	$CH_3(CH_2)_2CH_3$
pentane	C_5H_{12}	$CH_3(CH_2)_3CH_3$
hexane	C_6H_{14}	$CH_3(CH_2)_4CH_3$
general formula	C_nH_{2n+2}	

The alkanes

The **structural formula** shows how the atoms in the molecule are arranged. We can draw a **displayed structural formula** (sometimes called a full structural formula) which shows all the bonds. Here are the displayed structural formulae for the first three alkanes of the series:

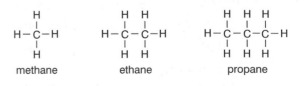

methane   ethane   propane

Alkanes do not take part in many chemical reactions. Their important reaction is combustion.

Alkanes contain only single bonds between carbon atoms. Such hydrocarbons are called **saturated hydrocarbons**. This is in contrast to the alkenes, which contain double bonds and are **unsaturated hydrocarbons**.

Isomers

Different compounds may have the same molecular formula. For example, butane and methylpropane are different alkanes which both have the molecular formula, C_4H_{10}. The structural formulae are:

butane $CH_3(CH_2)_2CH_3$; methylpropane $CH_3CH(CH_3)_2$.

The displayed structural formulae (or full structural formulae) showing every bond in each molecule are:

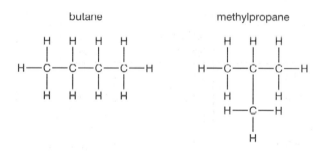

Compounds, such as these, with the same molecular formula and different structural formulae are **isomers**.

6.2 Cycloalkanes

Another homologous series of hydrocarbons is the **cycloalkanes**. They have the general formula C_nH_{2n}. Some members are:

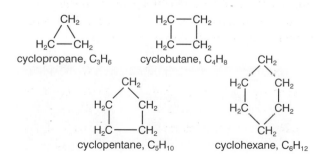

Cycloalkanes are isomeric with alkenes. The isomers of the cycloalkanes shown above are propene, butene, pentene and hexene.

6.3 Alkenes

The alkenes are a homologous series of hydrocarbons, as shown in the table.

name	molecular formula	structural formula
ethene	C_2H_4	$CH_2=CH_2$
propene	C_3H_6	$CH_3CH=CH_2$
butene	C_4H_8	$CH_3CH_2CH=CH_2$
pentene	C_5H_{10}	$CH_3(CH_2)_2CH-CH_2$
hexene	C_6H_{12}	$CH_3(CH_2)_3CH=CH_2$
general formula	C_nH_{2n}	

The alkenes

The double bond (see page 31) between the carbon atoms is the **functional group** of alkenes, and is responsible for their reactions. Alkenes are described as **unsaturated hydrocarbons**. They will react with hydrogen in an **addition reaction** to form saturated hydrocarbons (alkanes).

ethene + hydrogen → ethane

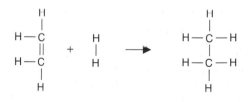

6.4 Reactions of alkenes

Alkenes are not used as fuels because they are an important source of other compounds. The double bond makes them chemically reactive, and they are starting materials in the manufacture of plastics, fibres, solvents and other chemicals.

Addition reactions

1 **Halogens add** to alkenes. A solution of bromine in an organic solvent is brown. If an alkene is bubbled through such a bromine solution, the

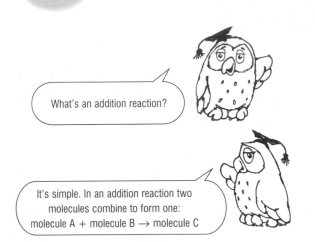

What's an addition reaction?

It's simple. In an addition reaction two molecules combine to form one:
molecule A + molecule B → molecule C

solution loses its colour. Bromine has added to the alkene to form a colourless compound, for example:

ethene + bromine → 1, 2-dibromoethane

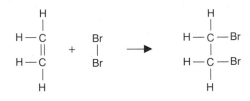

The decolorisation of a bromine solution is used to distinguish between an alkene and an alkane or cycloalkane.

2   **Hydrogenation**: animal fats, such as butter, are saturated compounds and are solid. Vegetable oils, such as sunflower seed oil, are unsaturated compounds and are liquid. An unsaturated vegetable oil can be converted into a saturated fat by hydrogenation (the addition of hydrogen):

vegetable oil + hydrogen $\xrightarrow[\text{nickel catalyst}]{\text{pass over heated}}$ solid fat
(unsaturated)   (saturated)

The solid fat produced is sold as margarine.

round-up

How much have you improved?
Work out your improvement index on page 111–112.

1 **a)** Write **(i)** the molecular formulae **(ii)** the structural formulae of ethane and ethene.   [4]
   b) State the difference between the chemical bonding in the two compounds.   [4]

2 **a)** Give two examples of substances which react with ethene but not with ethane.   [2]
   b) What name is given to the type of reactions in **a**? [1]
   c) How can ethene be converted into ethane?   [2]

3 Alkanes and alkenes burn in a similar way. Alkanes are important fuels.
   a) What are the main products of combustion of alkanes and alkenes?   [2]
   b) Why are alkenes not used as fuels?   [1]

4 **a)** Sketch the structural formulae of propane and propene.   [2]
   b) Which of the two compounds will decolorise bromine water?   [1]
   c) Sketch the structural formulae of the product formed in the reaction between bromine and the compound in **b**.   [1]

5 **a)** Write the structural formula of cyclopropane, C_3H_6 [1]
   b) What products are formed when cyclopropane burns in an excess of air?   [2]
   c) What reaction takes place between cyclopropane and bromine solution?   [1]
   d) Give the name and structural formula of an alkene which is an isomer of cyclopropane   [2]

6 Write the molecular formulae of:
   a) the simplest alkene
   b) pentane
   c) pentene
   d) hexane
   e) octene
   f) the alkene with 12 carbon atoms per molecule.   [6]

7 Write the structural formulae for the reactions **(i)** between butene and hydrogen **(ii)** between propene and bromine.   [4]

8 **a)** What is meant by a homologous series?   [4]
   b) When alkanes such as $C_{14}H_{30}$ are cracked, the products include alkenes as well as alkanes. Why is it not possible for a molecule of $C_{14}H_{30}$ to give 2 molecules of heptane?   [1]

The ionic bond

preview

At the end of this topic you will:

- understand how atoms combine by forming ionic bonds and covalent bonds

- understand how ionic compounds and covalent compounds differ in properties and structure.

- explain what happens when compounds are electrolysed

- give examples of the use of electrolysis in industry.

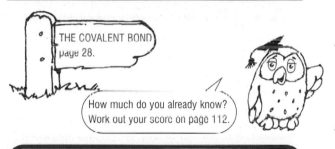

THE COVALENT BOND
page 28.

How much do you already know?
Work out your score on page 112.

Test yourself

1 Write words in the spaces to complete the sentences.

When an ionic compound is formed, some atoms lose electrons to become _____ ions while other atoms gain electrons to become _____ ions. Ions are held together by _____ attraction in a three-dimensional structure called a _____ . [4]

2 The element E has the electron arrangement E(2.8.2). The element Q has the electron arrangement Q(2.8.7). Explain what happens to atoms of E and Q when they combine to form an ionic compound, and give the formula of the compound. [3]

3 Name the following compounds:
a) $MgBr_2$ b) $FeCl_2$ c) $FeCl_3$ d) Na_2O e) $BaSO_4$. [5]

4 State the formulae of the following compounds:
a) potassium bromide   b) calcium carbonate
c) lead(II) oxide   d) lead(II) sulphate
e) silver chloride.   [5]

5 What do the electron arrangements of F^- and Na^+ and Ne have in common? Suggest another ion with the same electron arrangement as F^-. [2]

6 Name the particles which carry an electric current in
a) copper
b) potassium chloride solution
c) molten lead(II) bromide. [5]

7 Explain why
a) Potassium chloride conducts electricity in solution but not as as solid.
b) Lead(II) bromide conducts electricity when molten but not in the solid state.
c) Bromoethane, C_2H_5Br, does not conduct electricity. [6]

8 Which of the following are ionic compounds?
C_2H_6, N_2O, NH_3, CaF_2, H_2S, HF, KI [2]

9 Which of the following solids conduct electricity?
A zinc   B sulphur   C bronze
D crystalline copper(II) sulphate [2]

10 Which of the following liquids conduct electricity?
A a solution of dilute sulphuric acid
B a solution of sodium sulphate
C ethanol
D a solution of ethanoic acid [3]

11 What do the following terms mean?
a) an electrolyte b) an electrode [4, 1]

12 a) Explain the terms cation and anion, and give two examples of each. [6]
b) Explain why ions move towards electrodes. [2]
c) Explain why solid copper(II) chloride does not conduct electricity, but an aqueous solution of copper(II) chloride does. [2]

7.1 Bond formation

When chemical reactions take place, it is the electrons in the outer shell that are involved in the formation of bonds. The resistance of the noble gases to chemical change is believed to be due to the stability of the full outer shell of eight electrons (two for helium). When atoms react, they gain, lose or share electrons to attain an outer shell of eight electrons. Metallic elements frequently combine with non-metallic elements to form compounds.

7.2 Ionic bonding

Example 1

Sodium burns in chlorine to form sodium chloride. This is what happens to the electrons:

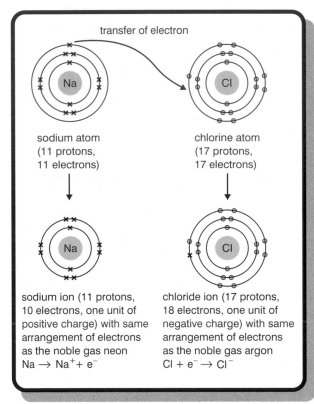

transfer of electron

sodium atom
(11 protons,
11 electrons)

chlorine atom
(17 protons,
17 electrons)

sodium ion (11 protons, 10 electrons, one unit of positive charge) with same arrangement of electrons as the noble gas neon
$Na \rightarrow Na^+ + e^-$

chloride ion (17 protons, 18 electrons, one unit of negative charge) with same arrangement of electrons as the noble gas argon
$Cl + e^- \rightarrow Cl^-$

The formation of sodium chloride

THE COVALENT
BOND
Page 28.

There is an electrostatic force of attraction between oppositely charged ions. This force is called an **ionic bond** or **electrovalent bond**. The ions Na^+ and Cl^- are part of a **giant ionic structure** (a crystal). The ions cannot move out of their positions in the structure, and the crystal cannot conduct electricity. When the solid is melted or dissolved, the ions become free to move and conduct electricity (see page 45).

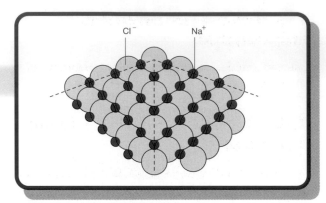

Cl^- Na^+

The structure of sodium chloride

Example 2

Magnesium + fluorine → magnesium fluoride

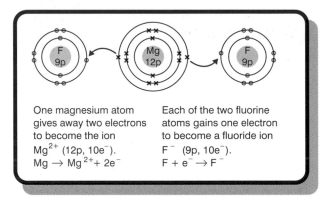

One magnesium atom gives away two electrons to become the ion Mg^{2+} (12p, 10e$^-$).
$Mg \rightarrow Mg^{2+} + 2e^-$

Each of the two fluorine atoms gains one electron to become a fluoride ion F^- (9p, 10e$^-$).
$F + e^- \rightarrow F^-$

The formula of magnesium fluoride is $Mg^{2+}(F^-)_2$ or MgF_2.

Example 3

Magnesium + oxygen → magnesium oxide

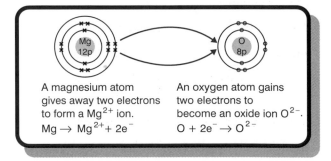

A magnesium atom gives away two electrons to form a Mg^{2+} ion.
$Mg \rightarrow Mg^{2+} + 2e^-$

An oxygen atom gains two electrons to become an oxide ion O^{2-}.
$O + 2e^- \rightarrow O^{2-}$

The formula of magnesium oxide is $Mg^{2+}O^{2-}$ or MgO.

7.3 Ionic and covalent substances

The concept map overleaf summarises the types of bonds formed by different substances.

7.4 The formula of an ionic compound

Ionic compounds consist of positive and negative ions. A sample of the compound is uncharged because the positive and negative charges balance exactly. In a sample of calcium chloride, $CaCl_2$, the number of chloride ions, Cl^-, is exactly twice the number of calcium ions, Ca^{2+}, so the formula is $Ca^{2+}(Cl^-)_2$.

Worked example

Compound:	*sodium sulphate*
Ions present are:	Na^+ and SO_4^{2-}
To balance the charges:	two Na^+ are needed to balance one SO_4^{2-}
The ions needed are:	$2Na^+$ and SO_4^{2-}
The formula is:	Na_2SO_4
Compound:	*iron(II) sulphate*
Ions present are:	Fe^{2+} and SO_4^{2}
To balance the charges:	one Fe^{2+} balances one SO_4^{2-}
The ions needed are:	Fe^{2+} and SO_4^{2-}
The formula is:	$FeSO_4$
Compound:	*iron(III) sulphate*
Ions present are:	Fe^{3+} and SO_4^{2-}
To balance the charges:	two Fe^{3+} balance three SO_4^{2-}
The ions needed are:	$2Fe^{3+}$ and $3SO_4^{2-}$
The formula is:	$Fe_2(SO_4)_3$

Note

The sulphates of iron are named iron(II) sulphate and iron(III) sulphate. The Roman numerals II and III show which type of ion, Fe^{2+} or Fe^{3+}, is present.

The symbols and charges of some ions

If you know the symbols and charges of these ions, you can work out the formulae of their compounds.

H^+ Na^+ K^+ Ag^+ NH_4^+	OH^- NO_3^- HCO_3^- Cl^- Br^- I^-
Ca^{2+} Cu^{2+} Zn^{2+} Pb^{2+} Mg^{2+} Fe^{2+} Ba^{2+}	SO_4^{2-} SO_3^{2-} CO_3^{2-} O^{2-} S^{2-}
Al^{3+} Fe^{3+}	

7.5 Conducting electricity

When substances conduct electricity changes happen at the electrodes. For example, a solution of copper chloride gives a deposit of copper at the negative electrode and a stream of chlorine at the positive electrode. The explanation is that copper chloride consists of positively charged particles of copper, called copper ions, and negatively charged particles of chlorine, called chloride ions. Experiments show that the copper ion carries two units of positive charge, Cu^{2+}, whereas the chloride ion carries one unit of negative charge, Cl^-. Copper chloride contains two chloride ions for every copper ion so that the charges balance, and the formula is $CuCl_2$.

In solid copper chloride, the ions are not free to move because they are held in a three-dimensional crystal structure, and the solid does not conduct electricity. When the salt is dissolved in water, the ions become free to move, the solution conducts electricity and electrolysis occurs.

There is another way of giving the ions freedom to move – to melt the solid. The electrolysis of molten sodium chloride is used for the extraction of sodium.

IONIC AND COVALENT SUBSTANCES

Ionic bonding
Ionic compounds are formed when a metallic element combines with a non-metallic element. An **ionic bond** is formed by **transfer of electrons** from one atom to another to form ions.

Covalent bonding
Atoms of non-metallic elements combine with other non-metallic elements by **sharing pairs of electrons** in their outer shells. A shared pair of electrons is a **covalent bond**.

There are three **types of covalent substances**.

1 Many covalent substances are composed of small individual molecules with only very small forces of attraction between molecules, e.g. the gases HCl, SO_2, CO_2, CH_4.

2 Some covalent substances consist of small molecules with weak forces of attraction between molecules, e.g. the volatile liquid ethanol, C_2H_5OH, and solid carbon dioxide.

3 Some covalent substances consist of giant molecules, e.g. quartz (silicon(IV) oxide). These substances have high melting and boiling points.

Atoms of **metallic elements** form positive ions (cations). Elements in Groups 1, 2 and 3 of the periodic table form ions with charges +1, +2 and +3, e.g. Na^+, Mg^{2+}, Al^{3+}. Atoms of **non-metallic elements** form negative ions (anions). Elements in Groups 6 and 7 of the periodic table form ions with charges –2 and –1, e.g. O^{2-} and Cl^-.

The maximum number of covalent bonds that an atom can form is equal to the number of electrons in the outer shell. An atom may not use all its outer electrons in bond formation.

Ionic compounds are **electrolytes** – they conduct electricity when molten or in solution and are split up (**electrolysed**) in the process. Covalent compounds are **non-electrolytes**

The strong electrostatic attraction between ions of opposite charge is an **ionic bond**. An ionic compound is composed of a giant regular structure of ions (see diagram of sodium chloride structure on page 30). This regular structure makes ionic compounds **crystalline**. The strong forces of attraction between ions make it difficult to separate the ions, and ionic compounds therefore have **high melting and boiling points**.

Organic solvents, e.g. ethanol and propanone, have covalent bonds. They dissolve covalent compounds but not ionic compounds.

Concept map: ionic and covalent substances

7.6 Ions

How is an ion formed from an atom? Atoms are uncharged. The number of protons in an atom is the same as the number of electrons. If an atom either gains or loses electrons, it will become electrically charged. Metal atoms and hydrogen atoms form positive ions (**cations**) by losing one or more electrons. Atoms of non-metallic elements form negative ions (**anions**) by gaining one or more electrons.

ATOMIC STRUCTURE Pages 24–25.

sodium atom Na → electron e⁻ + sodium ion Na⁺

(11 protons,   (11 protons,
11 electrons,   10 electrons,
charge = 0)   charge = +1)

chlorine atom Cl + electron e⁻ → chloride ion Cl⁻

(17 protons,   (17 protons,
17 electrons,   18 electrons,
charge = 0)   charge = −1)

The following table gives the symbols and formulae of some ions.

cations	anions
hydrogen ion H^+	bromide ion Br^-
sodium ion Na^+	chloride ion Cl^-
copper(II) ion Cu^{2+}	iodide ion I^-
lead(II) ion Pb^{2+}	hydroxide ion OH^-
aluminium ion Al^{3+}	nitrate ion NO_3^-
	sulphate ion SO_4^{2-}

7.7 At the electrodes

Copper(II) chloride solution

The diagram below shows what happens at the electrodes when copper(II) chloride is electrolysed.

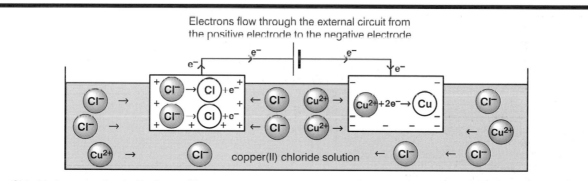

Chloride ions are attracted to the positive electrode. The positive charge enables the electrode to take electrons from chloride ions, discharging them to form chlorine atoms. Chlorine atoms then combine to form molecules. The electrode process is

chloride ion → chlorine atom + electron

$$Cl^-(aq) \rightarrow Cl(g) + e^-$$

followed by

$$2Cl(g) \rightarrow Cl_2(g)$$

Copper ions are attracted to the negative electrode. The negative charge on the electrode is due to the presence of electrons, and copper ions take electrons and are discharged to become copper atoms. The electrode process is

copper(II) ion + 2 electrons → copper atom

$$Cu^{2+}(aq) + 2e^- \rightarrow Cu(s)$$

The electrolysis of copper(II) chloride with carbon electrodes

7.8 Applications of electrolysis

Extraction of metals from their ores

★ **Sodium** is obtained from sodium chloride in the mineral rock salt. When molten anhydrous sodium chloride is electrolysed, the products are sodium and chlorine.

★ **Potassium**, **calcium** and **magnesium** are also obtained by electrolysis of molten anhydrous chlorides. The cost of electricity makes this method of extracting metals expensive.

★ **Aluminium** is mined as the ore bauxite, $Al_2O_3.2H_2O$. Purified anhydrous aluminium oxide is obtained from the ore. Before the oxide can be electrolysed, it must be melted. The high melting point of aluminium oxide, 2050°C, makes this difficult so it is dissolved in molten cryolite, Na_3AlF_6, at 1000°C before electrolysis.

Electroplating

Electrolysis is used to coat a metal with a thin even film of another metal, as shown in the diagram opposite.

1 A cheaper metal may be coated with a more beautiful and more expensive metal, e.g. silver or gold.

2 To prevent steel from rusting, it is electroplated with nickel and chromium which give the steel a bright surface that is not corroded in air.

3 Food cans are made of iron plated with a layer of tin. Tin is not corroded by food juices.

4 A layer of zinc is applied to iron in the manufacture of galvanised iron. Electroplating is one of the methods employed.

7.9 Coloured compounds

compound	colour of an aqueous solution	ion	deduce the colour of each aqueous ion
Sodium chloride	colourless	Sodium	
Sodium chromate	yellow	Potassium	
Potassium chloride	colourless	Copper(II)	
Potassium chromate	yellow	Iron(II)	
Copper(II) chloride	blue	Iron(III)	
Copper(II) nitrate	blue	Nickel	
Copper(II) sulphate	blue	Aluminium	(Answers on the next page).
Sodium sulphate	colourless	Cobalt(II)	
Potassium nitrate	colourless	Chloride	
Potassium permanganate	purple	Chromate	
Iron(II) sulphate	green	Nitrate	
Iron(III) sulphate	yellow-brown	Sulphate	
Nickel sulphate	green	Permanganate	
Aluminium sulphate	colourless		
Cobalt(II) chloride	blue		

Electroplating ▼

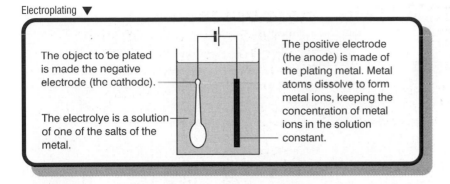

The object to be plated is made the negative electrode (the cathode).

The electrolye is a solution of one of the salts of the metal.

The positive electrode (the anode) is made of the plating metal. Metal atoms dissolve to form metal ions, keeping the concentration of metal ions in the solution constant.

Why does electricity make some substances move?

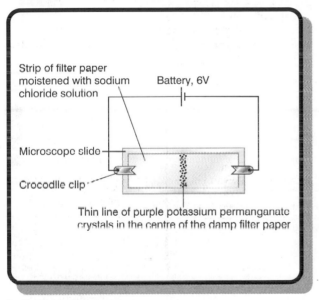

Strip of filter paper moistened with sodium chloride solution

Battery, 6V

Microscope slide

Crocodile clip

Thin line of purple potassium permanganate crystals in the centre of the damp filter paper

Looking for movement

Can you explain these observations?

a) When a current is passed for 15 minutes, the purple colour moves towards the positive electrode.

b) When the experiment is repeated using a thin line of blue crystals of copper(II) sulphate, the blue colour moves towards the negative electrode.

c) With potassium chromate, the yellow colour moves towards the positive electrode.

d) With potassium dichromate, the orange colour moves towards the positive electrode.

e) With iron(II) sulphate, the green colour moves towards the negative electrode. The ions present are:
  a) potassium permanganate, $KMnO_4$: K^+ and MnO_4^-;
  b) copper(II) sulphate, $CuSO_4$: Cu^{2+} and SO_4^{2-};
  c) potassium chromate, K_2CrO_4: K^+ and CrO_4^{2-};
  d) potassium dichromate, $K_2Cr_2O_7$: K^+ and $Cr_2O_7^{2-}$;
  f) iron(II) sulphate, $FeSO_4$: Fe^{2+} and SO_4^{2-}.

(Answers below.)

Answer to 'Coloured compounds'

Sodium, potassium, aluminium, chloride, nitrate, sulphate: colourless; copper(II): blue; cobalt(II): pink; iron(II) and nickel: green; iron (III): yellow-brown; chromate: yellow; permanganate: purple.

Answers to 'Why does electricity make some substances move?'

a) The purple colour is due to purple permanganate ions which are negatively charged. The potassium ions are colourless.

b) The blue colour is due to blue positively charged copper(II) ions. The sulphate ions are colourless.

c) The yellow chromate ions are negatively charged and move towards the positive electrode. The potassium ions are colourless.

d) The orange dichromate ions are negatively charged and move towards the positive electrode. The potassium ions are colourless.

e) The positively charged green iron(II) ions move towards the negative electrode. The sulphate ions are colourless.

round-up

How much have you improved?
Work out your improvement index on page 112–113.

1 Which of the substances in the table below could be
a) copper b) ethanol c) potassium chloride d) sucrose
e) mercury? [5]

substance	state at room temperature	does it conduct electricity?		
		solid state	liquid state	aqueous solution
A	s	no	yes	yes
B	s	yes	yes	insoluble
C	s	no	no	no
D	l		no	no
E	l		yes	insoluble

2 Coloured crystals were placed on a strip of filter paper
moistened with sodium chloride solution. A voltage was
applied across the paper. The diagram shows how
coloured streaks moved across the paper. What
explanation can you give? [4]

Crystal of copper(II)
sulphate (blue)

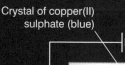

Crystal of potassium
permanganate (purple)

3 a) An aqueous solution of copper(II) chloride can be
electrolysed. Name the two products formed and say at
which electrode each is formed. Write ion–electron
equations for both electrode processes. [8]
b) Magnesium chloride, $MgCl_2$, is a solid of high melting
point and tetrachloromethane, CCl_4, is a volatile liquid.
Explain how differences in chemical bonding account
for these properties. [4]

4 Write formulae for any ions formed by the following
elements which have the same electron arrangement as
argon atoms: S, O, Cl, Ca, Si, Mg. [4]

5 How could you use the apparatus below to distinguish
between a) lead(II) nitrate solution b) dilute hydrochloric
acid c) ethanol solution? [4]

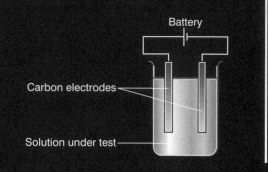

Battery

Carbon electrodes

Solution under test

6 Write formulae (see ions on page 47) for silver hydroxide,
copper(II) hydroxide, iron(II) hydroxide, iron (III) hydroxide,
ammonium bromide, ammonium sulphate, sodium nitrate,
sodium carbonate, aluminium chloride, aluminium
hydroxide, aluminium oxide, aluminium sulphate, calcium
hydroxide, calcium oxide and calcium sulphate. [15]

7 a) What are the particles in a crystal of
sodium chloride? [2]
b) What holds the particles together? [1]
c) Describe the arrangement of particles in
the crystal. [2]

8 Name the following compounds:
a) $Ca(OH)_2$ **b)** Na_2SO_3 **c)** $CuCO_3$
d) $Mg(HCO_3)_2$ **e)** KNO_3. [5]

9 State the formulae of the following compounds:
a) ammonium nitrate **b)** sodium sulphate
c) ammonium sulphate **d)** aluminium oxide
e) zinc hydroxide. [5]

Well done if you've improved. Don't worry
if you haven't. Take a break and try again.

Can you summarise this topic in a Mind Map?

Acids and bases

8

MIND MAP page 125.

preview

At the end of this section you will:

- **be able to define the terms 'acid', 'base' and 'alkali'**
- **calculate the relative mass formula of a compound**
- **understand 'the mole'.**

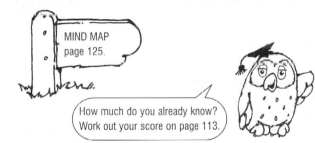

How much do you already know? Work out your score on page 113.

Test yourself

1 Say whether the substances listed are strongly acidic (SA), weakly acidic (WA), strongly basic (SB), weakly basic (WB) or neutral (N).
   a) battery acid, pH 0   [1]
   b) rainwater, pH 6.5   [1]
   c) blood, pH 7.4   [1]
   d) sea water, pH 8.5   [1]
   e) cabbage juice, pH 5.0   [1]
   f) saliva, pH 7.0   [1]
   g) washing soda, pH 11.5   [1]

2 You are given two bottles labelled 'acid 1' and 'acid 2'. One is a weak acid and the other is a strong acid. Describe two tests you could do to find out which is which.   [3]

3 Name
   a) a strong acid present in your stomach [1]
   b) a base present in indigestion tablets [1]
   c) a weak acid present in fruits   [1]
   d) a weak base used as a domestic cleaning fluid. [1]

4 Kleenit is an oven spray for cleaning greasy ovens. It contains a concentrated solution of sodium hydroxide.
   a) Why does sodium hydroxide remove grease? [1]
   b) Why does sodium hydroxide work better than ammonia?   [1]
   c) What two safety precautions should you take when using Kleenit?   [2]
   d) Why does Moppit, a fluid used for cleaning floors, contain ammonia rather than sodium hydroxide? [1]
   e) Why do soap manufacturers use sodium hydroxide, not ammonia?   [1]

5 State the relative formula masses of a) SO_2 b) SO_3 c) H_2SO_4 d) CH_3CO_2H.   [5]

6 State the amount in moles of a) sodium in 46 g of sodium b) sulphur atoms in 64 g of sulphur c) S_8 molecules in 64 g of sulphur.   [3]

7 State the mass of a) mercury in 0.100 mol of mercury b) sulphuric acid in 0.25 mol of H_2SO_4 c) magnesium oxide in 3.0 mol of MgO.   [3]

8.1 Acids

Where are acids found?

The following are strong acids:

- Hydrochloric acid occurs in the stomach, where it aids digestion.
- Nitric acid is used in the production of fertilisers and explosives.
- Sulphuric acid is used car batteries and in the production of fertilisers.

The following are weak acids:

- Carbonic acid is used in fizzy drinks.
- Citric acid occurs in lemons and other citrus fruits.
- Ethanoic acid occurs in vinegar.
- Lactic acid is present in sour milk.

What do acids do?

Acids are compounds that release hydrogen ions when dissolved in water. The hydrogen ions are responsible for the typical reactions of acids.

The concept map on page 52 gives a summary of the properties of acids.

8.2 Bases

Where do you find bases?

The following are strong bases:

- Calcium hydroxide is used to treat soil which is too acidic.
- Calcium oxide is used in the manufacture of cement and concrete.
- Magnesium hydroxide is used in anti-acid indigestion tablets.
- Sodium hydroxide is used in soap manufacture and as a degreasing agent.
- The weak base ammonia is used in cleaning fluids, as a degreasing agent and in the manufacture of fertilisers.

What do bases do?

A **base** is a substance that reacts with an acid to form a salt and water as the only products. A soluble base is called an **alkali**. Sodium hydroxide, NaOH, is a strong base and a strong alkali. Its reactions are less vigorous than those of sodium hydroxide.

BASES
Page 61.

The concept map opposite gives a summary of the properties of bases.

Oxides

The oxides of metallic elements are basic. The oxides of non-metallic elements are acidic or neutral; see the concept map below.

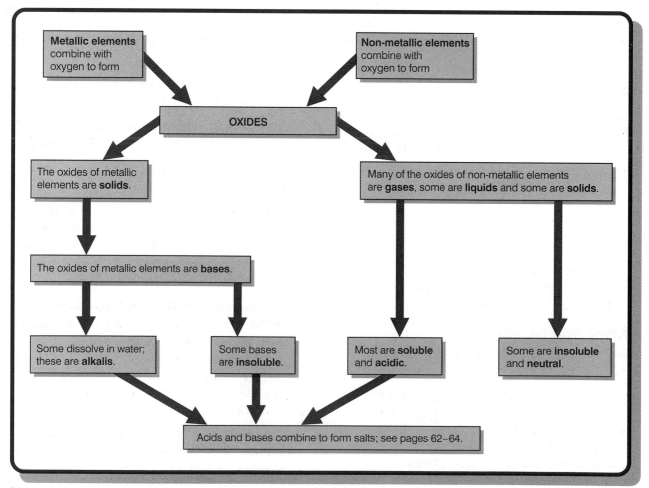

Concept map: properties of oxides

Metallic elements combine with oxygen to form

Non-metallic elements combine with oxygen to form

OXIDES

The oxides of metallic elements are **solids**.

Many of the oxides of non-metallic elements are **gases**, some are **liquids** and some are **solids**.

The oxides of metallic elements are **bases**.

Some dissolve in water; these are **alkalis**.

Some bases are **insoluble**.

Most are **soluble** and **acidic**.

Some are **insoluble** and **neutral**.

Acids and bases combine to form salts; see pages 62–64.

8.3 pH

In **neutral** solutions, e.g. pure water, the concentration of hydrogen ions is equal to the concentration of hydroxide ions.

In **acidic** solutions the concentration of hydrogen ions is greater than the concentration of hydroxide ions.

In **alkaline** solutions the concentration of hydroxide ions is greater than the concentration of hydrogen ions.

The term **pH** is related to the concentration of hydrogen ions in a solution. The degree of acidity or alkalinity of a solution is measured by its pH value:

- a neutral solution has a pH of 7
- an acidic solution has pH between 0 and 7
- an alkaline solution has pH between 7 and 14.

Indicators

Indicators are substances which change colour in acidic and alkaline solutions. Universal indicator can distinguish between strongly acidic and weakly acidic solutions and between strongly alkaline and weakly alkaline solutions, as shown in the diagram below.

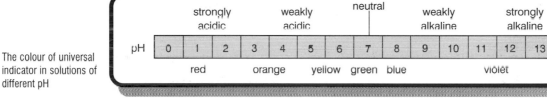

The colour of universal indicator in solutions of different pH

8.4 Acid rain

Rain is naturally weakly acidic because it dissolves carbon dioxide from the air. The pH of natural rainwater is 5.2. Rain with a pH below this is described as **acid rain**.

There are many effects of acid rain:

- damage to lakes and the fish and plants in them; see diagram on opposite page
- washing of nutrients out of topsoil, resulting in poor crops and damage to trees
- costly damage to building materials, e.g. limestone, concrete, cement and metal.

What can be done?

Sweden has tackled the problem by spraying tonnes of calcium hydroxide into acid lakes.

Members of the European Community (EC) have agreed to make a 60% to 70% cut in their emissions of sulphur dioxide by 2003. Power stations must

POLLUTION OF AIR Pages 36–37.

make a big contribution to solving the problem. Some lines of attack are:

1 Coal can be crushed and washed with a solvent to remove much of the sulphur content.

2 Fuel oil can be purified at the refinery – at a cost.

3 Sulphur dioxide can be removed from the exhaust gases of power stations. In **flue gas desulphurisation** (**FGD**), jets of wet powdered limestone neutralise acidic gases as they pass up the chimney of the power station.

4 In a **pulverised fluidised bed combustion** (**PFBC**) furnace, the coal is pulverised (broken into small pieces) and burnt on a bed of powdered limestone, which is 'fluidised' (kept in motion by an upward flow of air). As the coal burns, sulphur dioxide reacts with the limestone.

5 Nuclear power stations do not send pollutants into the air. However, they create the problem of storing radioactive waste.

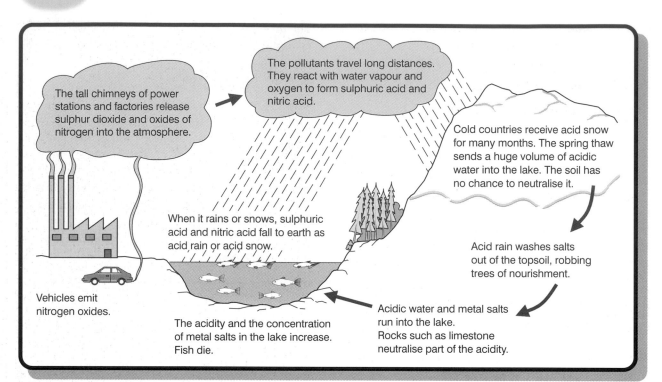

The tall chimneys of power stations and factories release sulphur dioxide and oxides of nitrogen into the atmosphere.

The pollutants travel long distances. They react with water vapour and oxygen to form sulphuric acid and nitric acid.

Cold countries receive acid snow for many months. The spring thaw sends a huge volume of acidic water into the lake. The soil has no chance to neutralise it.

When it rains or snows, sulphuric acid and nitric acid fall to earth as acid rain or acid snow.

Acid rain washes salts out of the topsoil, robbing trees of nourishment.

Vehicles emit nitrogen oxides.

The acidity and the concentration of metal salts in the lake increase. Fish die.

Acidic water and metal salts run into the lake. Rocks such as limestone neutralise part of the acidity.

Acid rain; its source and its effect on lake water

8.5 Some definitions

Relative atomic mass

Relative atomic mass RAM of element =

$$\frac{\text{mass of one atom of the element}}{\frac{1}{12} \text{ mass of one atom of carbon-12}}$$

Relative formula mass

Relative formula mass RFM of compound =

$$\frac{\text{mass of one molecule or formula unit}}{\frac{1}{12} \text{ mass of one atom of carbon-12}}$$

The relative formula mass RFM of a compound is the sum of the relative atomic masses of all the atoms in a molecule or formula unit of the compound. For sulphuric acid, H_2SO_4,

RFM = [2 × RAM (H)] + RAM(S) + [4 × RAM(O)]

8.6 The mole

Chemists know that in a reaction

A + B → C

n atoms or molecules of A react with n atoms or molecules of B to form n molecules of C. Chemists may want to provide equal numbers of atoms or molecules of A and B to react together. How can they do this? Thanks to the **mole concept** they can *count out* atoms and molecules by *weighing* A and B. The mole concept makes the connection between the *mass* of a quantity substance and the *number* of atoms or molecules present.

★ One mole of each and every element contains the same number of atoms (6×10^{23} atoms).

★ One mole of an element has a mass equal to the RAM expressed in grams.
Sodium has RAM = 23, and one mole of sodium has a mass of 23 g.

★ For every compound one mole of the compound contains the same number (6×10^{23}) of molecules or formula units.

★ One mole of a compound has a mass equal to the RFM expressed in grams.

Calcium hydroxide $Ca(OH)_2$ has RFM = 74, and one mole of $Ca(OH)_2$ has mass = 74 g.

Number of moles of an element =

$$\frac{\text{Mass of the element (g)}}{\text{Mass of one mole of the element (g)}}$$
(RAM of element in grams)

Number of moles of a compound =

$$\frac{\text{Mass of the compound (g)}}{\text{Mass of one mole of the compound (g)}}$$
(RFM of compound in grams)

Example 1

Find the number of moles of calcium present in 120 g of calcium.

RAM of Ca = 40 therefore mass of one mole of Ca = 40 g

No of moles of Ca = mass (grams) / RAM expressed in grams
= 120 g / 40 g = 3.0

There are 3.0 mol calcium in 120 g calcium. Note that mol is the abbreviation for mole or moles.

Example 2

If you need 2.25 mol magnesium carbonate, what mass of the substance do you weigh out?

RFM of $MgCO_3$ = 24 + 12 + (3 × 16) = 84

mass of substance (g) = no of moles × mass of one mole of substance (g)
= 2.25 × 84 g = 189 g

The mass of magnesium carbonate is 189 g.

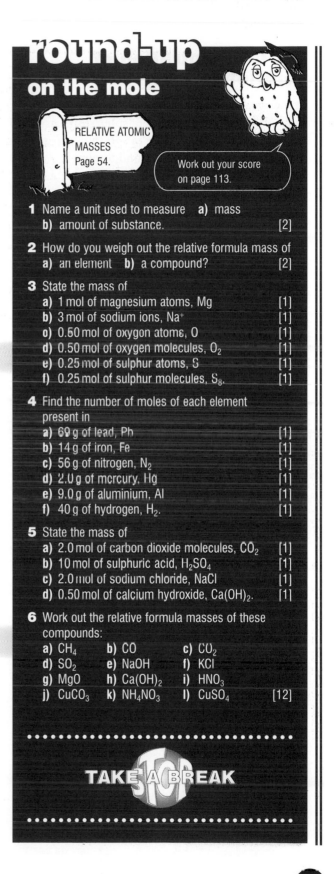

round-up
on the mole

RELATIVE ATOMIC MASSES Page 54.

Work out your score on page 113.

1 Name a unit used to measure a) mass b) amount of substance. [2]

2 How do you weigh out the relative formula mass of a) an element b) a compound? [2]

3 State the mass of
a) 1 mol of magnesium atoms, Mg [1]
b) 3 mol of sodium ions, Na^+ [1]
c) 0.50 mol of oxygen atoms, O [1]
d) 0.50 mol of oxygen molecules, O_2 [1]
e) 0.25 mol of sulphur atoms, S [1]
f) 0.25 mol of sulphur molecules, S_8. [1]

4 Find the number of moles of each element present in
a) 69 g of lead, Pb [1]
b) 14 g of iron, Fe [1]
c) 56 g of nitrogen, N_2 [1]
d) 2.0 g of mercury, Hg [1]
e) 9.0 g of aluminium, Al [1]
f) 40 g of hydrogen, H_2. [1]

5 State the mass of
a) 2.0 mol of carbon dioxide molecules, CO_2 [1]
b) 10 mol of sulphuric acid, H_2SO_4 [1]
c) 2.0 mol of sodium chloride, NaCl [1]
d) 0.50 mol of calcium hydroxide, $Ca(OH)_2$. [1]

6 Work out the relative formula masses of these compounds:
a) CH_4 b) CO c) CO_2
d) SO_2 e) NaOH f) KCl
g) MgO h) $Ca(OH)_2$ i) HNO_3
j) $CuCO_3$ k) NH_4NO_3 l) $CuSO_4$ [12]

TAKE A BREAK

55

8.7 Concentration of solution

One way of describing the concentration of a solution is to state the mass of solute (dissolved substance) present per litre of solution, e.g. grams per litre, g/l. Chemists often find it more convenient to state the number of *moles* of a solute present per litre of solution.

$$\frac{\text{Number of}}{\text{moles of solute}} = \frac{\text{Number of moles of solute}}{\text{Volume of solution in litres}}$$
per litre

Rearranging,

Number of   =   Volume of solution
moles of solute   (l) × Concentration (mol/l)

Example 1

Calculate the number of moles of solute present in 250 cm³ of a solution of hydrochloric acid which has a concentration of 2.0 mol/l .

Number of moles = Volume (l) × Concentration
(mol/l)

Number of   = 250 × 10⁻³ l × 2.0 mol/l
moles of HCl   = 0.50

The solution contains 0.50 mol of HCl.

Note: When you are given the volume in cm³ , you have to change it into litres. 1000 cm³ = 1 litre.

Example 2

What mass of sodium carbonate must be dissolved in 1 l of solution to give a solution of concentration 1.5 mol/l?

Number of moles = Volume (l) × Concentration
(mol/l)
= 1.00 l × 1.5 mol/l
= 1.5

RFM of sodium carbonate, Na_2CO_3 = (2 × 23) + 12 + (3 × 16)= 106

One mole of Na_2CO_3 has mass = 106 g

Mass of sodium carbonate = 1.5 × 106 g = 159 g

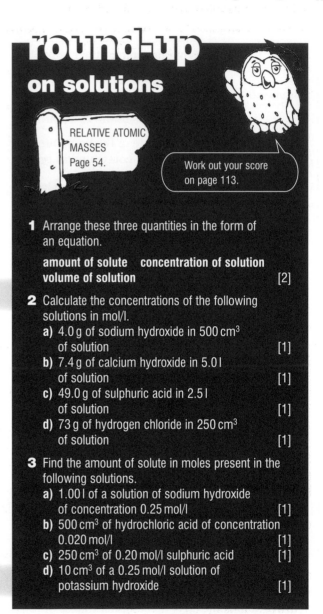

round-up

on solutions

RELATIVE ATOMIC MASSES Page 54.

Work out your score on page 113.

1 Arrange these three quantities in the form of an equation.

amount of solute   concentration of solution volume of solution [2]

2 Calculate the concentrations of the following solutions in mol/l.
a) 4.0 g of sodium hydroxide in 500 cm³ of solution [1]
b) 7.4 g of calcium hydroxide in 5.0 l of solution [1]
c) 49.0 g of sulphuric acid in 2.5 l of solution [1]
d) 73 g of hydrogen chloride in 250 cm³ of solution [1]

3 Find the amount of solute in moles present in the following solutions.
a) 1.00 l of a solution of sodium hydroxide of concentration 0.25 mol/l [1]
b) 500 cm³ of hydrochloric acid of concentration 0.020 mol/l [1]
c) 250 cm³ of 0.20 mol/l sulphuric acid [1]
d) 10 cm³ of a 0.25 mol/l solution of potassium hydroxide [1]

8.8 Concentration and pH

A solution of hydrochloric acid which contains 0.1 mol of hydrogen chloride per litre has pH of 1. If you dilute this solution ten times to get a solution containing 0.01 mol/l of hydrogen chloride the pH becomes 2. If you dilute the new solution ten times to get a solution containing 0.001 mol/l of hydrogen chloride the pH becomes 3. A ten-fold dilution increases the pH by 1 unit.

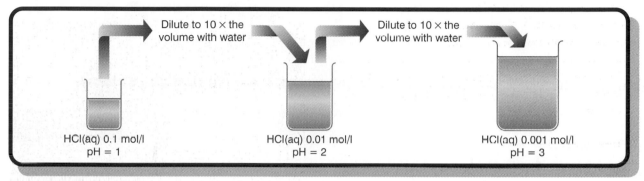

Concentration and pH of an acid

Acidity is caused by hydrogen ions. When you dilute the solution by a factor of ten, you are reducing the concentration of hydrogen ions by a factor of ten. As the pH increases, you will notice a decrease in the vigour of the chemical reactions of the acid.

A solution of sodium hydroxide which contains 0.1 mol of sodium hydroxide per litre has a pH of 13. A ten-fold dilution decreases the pH by 1 unit.

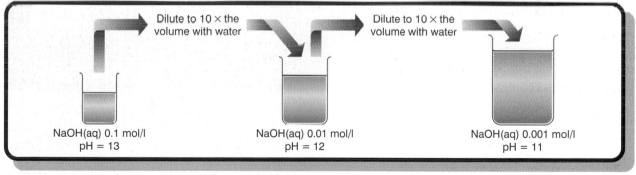

Concentration and pH of an alkali

When you dilute the alkaline solution by a factor of ten, you are reducing the concentration of hydroxide ions by a factor of ten. As the pH decreases, you will notice a decrease in the chemical reactivity of the alkali.

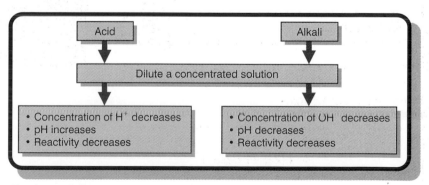

pH and concentration

round-up

on acids and bases

How much have you improved?
Work out your improvement index on page 113.

1 Write the chemical equations for the reactions with water of ammonia, calcium oxide and hydrogen chloride. [6]

2 Water is added to a dilute solution of sulphuric acid.

a) Does the solution become more concentrated or more dilute?

b) Does the pH increase or decrease?

c) Do the chemical reactions of the acid become more or less vigorous?

d) Does the concentration of hydrogen ions increase or decrease? [4]

3 a) In an acidic solution is the concentration of hydrogen ions **(i)** greater than **(ii)** less than **(iii)** equal to that in water? [1]

b) In an alkaline solution is the concentration of hydroxide ions **(i)** greater than **(ii)** less than or **(iii)** equal to that in water? [1]

c) In a neutral solution, is the concentration of hydrogen ions **(i)** greater than **(ii)** less than or **(iii)** equal to that in water?

4 Choose the statements which apply to
i) 1 l of dilute sulphuric acid
ii) 1 l of sodium hydroxide solution.
A contains more hydrogen ions than 1 l of pure water
B contains the same number of hydrogen ions of 1 l pure water
C contains more hydrogen ions than hydroxide ions
D contains more hydroxide ions than hydrogen ions
E contains the same number of hydroxide ions as 1 l of pure water
F contains more hydroxide ions than 1 l of pure water.

5 Select the correct answer or answers. $20 \, cm^3$ of 2.0 mol/l hydrochloric acid, HCl(aq), is neutralised by:
A $20 \, cm^3$ of 2.0 mol/l NaOH(aq)
B $20 \, cm^3$ of 1.0 mol/l NaOH(aq)
C $20 \, cm^3$ of 1.0 mol/l $Ca(OH)_2$(aq)
D $20 \, cm^3$ of 2.0 mol/l $Ca(OH)_2$(aq). [2]

Well done if you've improved. Don't worry if you haven't. Take a break and try again.

Reactions of acids

preview

At the end of this topic you will:

- **know the typical reactions of acids and bases**
- **know methods of preparing salts**
- **be able to do calculations on titrations.**

MIND MAP
page 124.

How much do you already know?
Work out your score on page 114.

Test yourself

1 Copy and complete the following word equations.
   a) hydrochloric acid + calcium oxide →
   b) hydrochloric acid + sodium carbonate →
   c) sulphuric acid + sodium hydroxide →
   d) sulphuric acid + copper(II) carbonate →
   e) sulphuric acid + magnesium →
   f) nitric acid + calcium hydroxide →   [14]

2 Copy and complete the following equations.
   a) $Mg(s) + 2HCl(aq) \rightarrow H_2(g) + \underline{\quad}$
   b) $MgO(s) + H_2SO_4(aq) \rightarrow \underline{\quad} + \underline{\quad}$
   c) $CaCO_3(s) + 2HCl(aq) \rightarrow \underline{\quad} + \underline{\quad} + \underline{\quad}$
   d) $2KOH(aq) + H_2SO_4(aq) \rightarrow \underline{\quad} + \underline{\quad}$ [8]

3 Describe how you could prepare:
   a) crystals of sodium nitrate from solutions of sodium hydroxide and dilute nitric acid   [3]
   b) crystals of cobalt(II) sulphate from cobalt(II) oxide and dilute sulphuric acid.   [3]

4 a) Sodium hydroxide and nitric acid react to form sodium nitrate and water. Write an ionic equation for the formation of water molecules.   [3]
   b) What name is given to the sodium ions and nitrate ions in the reaction?   [1]
   c) Sodium carbonate and hydrochloric acid react to form sodium chloride + carbon dioxide + water. Write an ionic equation for the formation of water and carbon dioxide.   [4]

5 Outline a method of preparation for the insoluble salt strontium carbonate.   [4]

9.1 The properties of acids

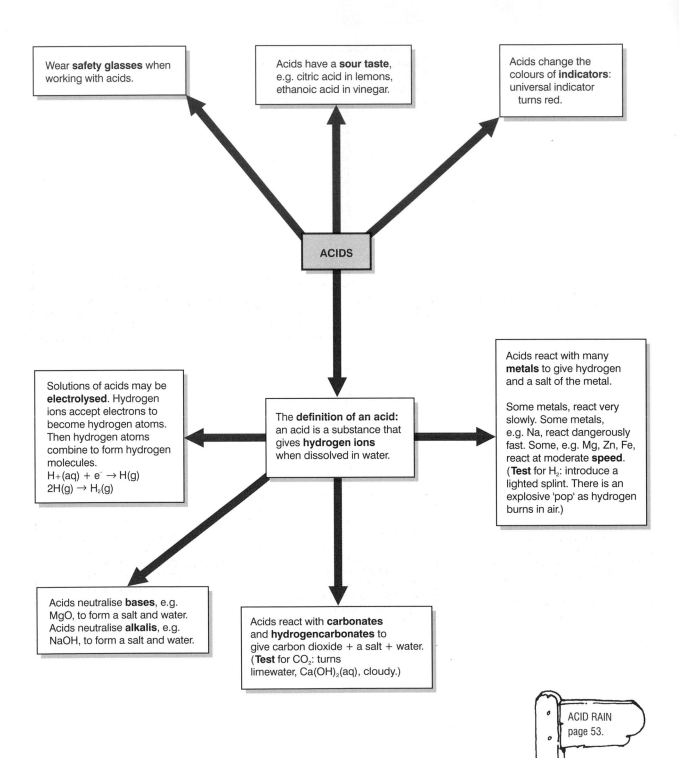

Wear **safety glasses** when working with acids.

Acids have a **sour taste**, e.g. citric acid in lemons, ethanoic acid in vinegar.

Acids change the colours of **indicators**: universal indicator turns red.

ACIDS

The **definition of an acid**: an acid is a substance that gives **hydrogen ions** when dissolved in water.

Solutions of acids may be **electrolysed**. Hydrogen ions accept electrons to become hydrogen atoms. Then hydrogen atoms combine to form hydrogen molecules.
$H_+(aq) + e^- \rightarrow H(g)$
$2H(g) \rightarrow H_2(g)$

Acids react with many **metals** to give hydrogen and a salt of the metal.

Some metals, react very slowly. Some metals, e.g. Na, react dangerously fast. Some, e.g. Mg, Zn, Fe, react at moderate **speed**. (**Test** for H_2: introduce a lighted splint. There is an explosive 'pop' as hydrogen burns in air.)

Acids neutralise **bases**, e.g. MgO, to form a salt and water. Acids neutralise **alkalis**, e.g. NaOH, to form a salt and water.

Acids react with **carbonates** and **hydrogencarbonates** to give carbon dioxide + a salt + water. (**Test** for CO_2: turns limewater, $Ca(OH)_2(aq)$, cloudy.)

ACID RAIN page 53.

Concept map: The properties of acids

9.2 The properties of bases

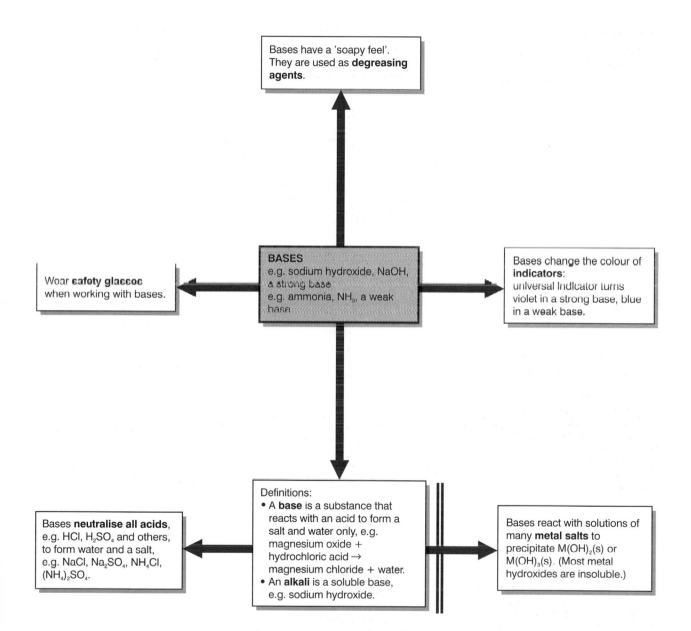

Bases have a 'soapy feel'. They are used as **degreasing agents**.

BASES
e.g. sodium hydroxide, NaOH, a strong base
e.g. ammonia, NH_3, a weak base

Wear **safety glasses** when working with bases.

Bases change the colour of **indicators**:
universal indicator turns violet in a strong base, blue in a weak base.

Bases **neutralise all acids**, e.g. HCl, H_2SO_4 and others, to form water and a salt, e.g. NaCl, Na_2SO_4, NH_4Cl, $(NH_4)_2SO_4$.

Definitions:
• A **base** is a substance that reacts with an acid to form a salt and water only, e.g. magnesium oxide + hydrochloric acid → magnesium chloride + water.
• An **alkali** is a soluble base, e.g. sodium hydroxide.

Bases react with solutions of many **metal salts** to precipitate $M(OH)_2(s)$ or $M(OH)_3(s)$. (Most metal hydroxides are insoluble.)

Concept map: The properties of bases

61

9.3 Equations for some reactions in the concept maps

Acids

With metals

zinc + sulphuric acid → hydrogen + zinc sulphate

$Zn(s) + H_2SO_4(aq) \rightarrow H_2(g) + ZnSO_4(aq)$

or the ionic equation:
$Zn(s) + 2H^+(aq) \rightarrow Zn^{2+}(aq) + H_2(g)$

With bases

magnesium oxide + sulphuric acid → magnesium sulphate + water

$MgO(s) + H_2SO_4(aq) \rightarrow MgSO_4(aq) + H_2O(l)$

or the ionic equation:
$O^{2-}(s) + 2H^+(aq) \rightarrow H_2O(l)$

calcium hydroxide + hydrochloric acid → calcium chloride + water

$Ca(OH)_2(s) + 2HCl(aq) \rightarrow CaCl_2(aq) + 2H_2O(l)$

or the ionic equation:
$OH^-(aq) + H^+(aq) \rightarrow H_2O(l)$

With carbonates

calcium carbonate + hydrochloric acid → carbon dioxide + calcium chloride + water

$CaCO_3(s) + 2HCl(aq) \rightarrow CO_2(g) + CaCl_2(aq) + H_2O(l)$

or the ionic equation:
$CO_3^{2-}(s) + 2H^+(aq) \rightarrow CO_2(g) + H_2O(l)$

With alkalis

ammonia + sulphuric acid → ammonium sulphate

$2NH_3(aq) + H_2SO_4(aq) \rightarrow (NH_4)_2SO_4(aq)$

Bases

With acids

magnesium oxide + hydrochloric acid → magnesium chloride + water

$MgO(s) + 2HCl(aq) \rightarrow MgCl_2(aq) + H_2O(l)$

or the ionic equation:
$O^{2-}(s) + 2H^+(aq) \rightarrow H_2O(l)$

sodium hydroxide + hydrochloric acid → sodium chloride + water

$NaOH(aq) + HCl(aq) \rightarrow NaCl(aq) + H_2O(l)$

sodium hydroxide + sulphuric acid → sodium sulphate + water

$2NaOH(aq) + H_2SO_4(aq) \rightarrow Na_2SO_4(aq) + 2H_2O(l)$

or the ionic equation:
$OH^-(aq) + H^+(aq) \rightarrow H_2O(l)$

With metal salts

sodium hydroxide + iron(II) sulphate → iron(II) hydroxide + sodium sulphate

$2NaOH(aq) + FeSO_4(aq) \rightarrow Fe(OH)_2(s) + Na_2SO_4(aq)$

or the ionic equation:
$Fe^{2+}(aq) + 2OH^-(aq) \rightarrow Fe(OH)_2(s)$

sodium hydroxide + iron(III) sulphate → iron(III) hydroxide + sodium sulphate

$6NaOH(aq) + Fe_2(SO_4)_3(aq) \rightarrow 2Fe(OH)_3(s) + 3Na_2SO_4(aq)$

or the ionic equation:
$Fe^{3+}(aq) + 3OH^-(aq) \rightarrow Fe(OH)_3(s)$

9.4 Neutralisation

Neutralisation is the combination of hydrogen ions (from an acid) and hydroxide ions (from an alkali) or oxide ions (from an insoluble base) to form water. In the process a salt is formed. For example, with an alkali:

hydrochloric acid + sodium hydroxide → sodium chloride + water

$HCl(aq) + NaOH(aq) \rightarrow NaCl(aq) + H_2O(l)$

acid + alkali → salt + water

The hydrogen ions and hydroxide ions combine to form water molecules.

$$H^+(aq) + OH^-(aq) \rightarrow H_2O(l)$$

Sodium ions and chloride ions remain in the solution, which becomes a solution of sodium chloride.

With a base:

$$\text{sulphuric acid} + \text{copper(II) oxide} \rightarrow \text{copper(II) sulphate} + \text{water}$$

$$H_2SO_4(aq) + CuO(s) \rightarrow CuSO_4(aq) + H_2O(l)$$

acid + base → salt + water

Hydrogen ions and oxide ions combine to form water:

$$2H^+(aq) + O^{2-}(s) \rightarrow H_2O(l)$$

Copper(II) ions and sulphate ions remain in the solution, which becomes a solution of copper(II) sulphate.

9.5 Salts

A salt is a compound in which the hydrogen ions of an acid have been replaced by metal ions or ammonium ions, e.g. sodium chloride NaCl, calcium chloride $CaCl_2$, copper sulphate $CuSO_4$, sodium carbonate Na_2CO_3, sodium hydrogencarbonate $NaHCO_3$, ammonium chloride NH_4Cl.

Making soluble salts

To make a soluble salt, an acid is neutralised by adding a metal, a solid base, a solid metal carbonate or a solution of an alkali.

★ **Method 1:** acid + metal → salt + hydrogen

★ **Method 2:** acid + metal oxide → salt + water

★ **Method 3:**

acid + metal carbonate → salt + water + carbon dioxide

★ **Method 4:** acid + alkali → salt + water

Here are the practical details for methods **1**, **2** and **3**.

a) Carry out the neutralisation as shown in the table and diagram below.

b) Filter to remove the excess of solid, using a filter funnel and filter paper.

c) Evaporate the filtrate, preferably on a water bath.

d) As the solution cools, crystals of the salt form. Separate the crystals by filtration. Using a little distilled water, wash the crystals in the filter funnel. Leave the crystals to dry.

method 1 (acid + metal)	method 2 (acid + metal oxide)	method 3 (acid + metal carbonate)
Warm the acid, then switch off the Bunsen burner.	Warm the acid.	
Add an excess of the metal to the acid. When no more hydrogen is evolved, the reaction is complete.	Add an excess of the metal oxide to the acid. When the solution no longer turns pH paper red, the reaction is complete.	Add an excess of the metal carbonate to the acid. When no more carbon dioxide is evolved, the reaction is complete.

Neutralisation details for methods **1** to **3**

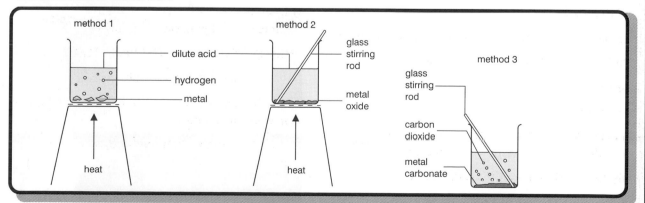

Adding an excess of a solid reactant to an acid

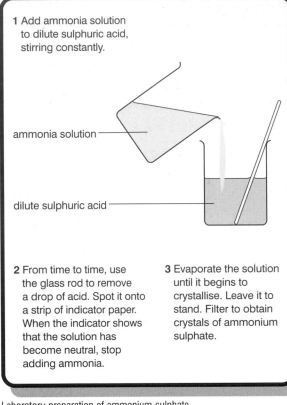

1 Add ammonia solution to dilute sulphuric acid, stirring constantly.

ammonia solution

dilute sulphuric acid

2 From time to time, use the glass rod to remove a drop of acid. Spot it onto a strip of indicator paper. When the indicator shows that the solution has become neutral, stop adding ammonia.

3 Evaporate the solution until it begins to crystallise. Leave it to stand. Filter to obtain crystals of ammonium sulphate.

Laboratory preparation of ammonium sulphate

Here are the practical details for method **4** (acid + alkali). For example, the neutralisation of an acid by ammonia solution can be used to make ammonium salts. The diagram above shows the steps in the preparation of ammonium sulphate.

Making insoluble salts: precipitation

Insoluble salts are made by mixing two solutions. For example, barium sulphate is insoluble. A soluble barium salt and a soluble sulphate must be chosen to make barium sulphate. The precipitate is separated by filtering or centrifuging.

The equation for the reaction is:

barium chloride + sodium sulphate → barium sulphate + sodium chloride

$$BaCl_2(aq) + Na_2SO_4(aq) \rightarrow BaSO_4(s) + 2NaCl(aq)$$

or the ionic equation:

$$Ba^{2+}(aq) + SO_4^{2-}(aq) \rightarrow BaSO_4(s)$$

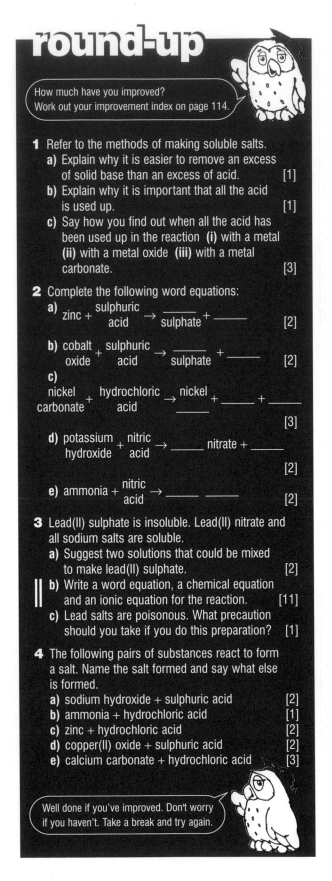

round-up

How much have you improved? Work out your improvement index on page 114.

1 Refer to the methods of making soluble salts.
  a) Explain why it is easier to remove an excess of solid base than an excess of acid. [1]
  b) Explain why it is important that all the acid is used up. [1]
  c) Say how you find out when all the acid has been used up in the reaction **(i)** with a metal **(ii)** with a metal oxide **(iii)** with a metal carbonate. [3]

2 Complete the following word equations:
  a) zinc + sulphuric acid → _____ sulphate + _____ [2]
  b) cobalt oxide + sulphuric acid → _____ sulphate + _____ [2]
  c) nickel carbonate + hydrochloric acid → nickel _____ + _____ + _____ [3]
  d) potassium hydroxide + nitric acid → _____ nitrate + _____ [2]
  e) ammonia + nitric acid → _____ _____ [2]

3 Lead(II) sulphate is insoluble. Lead(II) nitrate and all sodium salts are soluble.
  a) Suggest two solutions that could be mixed to make lead(II) sulphate. [2]
  b) Write a word equation, a chemical equation and an ionic equation for the reaction. [11]
  c) Lead salts are poisonous. What precaution should you take if you do this preparation? [1]

4 The following pairs of substances react to form a salt. Name the salt formed and say what else is formed.
  a) sodium hydroxide + sulphuric acid [2]
  b) ammonia + hydrochloric acid [1]
  c) zinc + hydrochloric acid [2]
  d) copper(II) oxide + sulphuric acid [2]
  e) calcium carbonate + hydrochloric acid [3]

Well done if you've improved. Don't worry if you haven't. Take a break and try again.

9.6 Titration

Example 1

In a titration of hydrochloric acid against sodium hydroxide solution, 15.0 cm³ of hydrochloric acid neutralise 25.0 cm³ of a 0.100 mol/l solution of sodium hydroxide. What is the concentration of the hydrochloric acid?

1 First write the equation for the reaction:

$$\text{hydrochloric} \atop \text{acid} \; + \; \text{sodium} \atop \text{hydroxide} \; \rightarrow \; \text{sodium} \atop \text{chloride} \; + \; \text{water}$$

$$HCl(aq) + NaOH(aq) \rightarrow NaCl(aq) + H_2O(l)$$

The equation tells you that 1 mol of HCl neutralises 1 mol of NaOH.

2 Now work out the number of moles of base. You must start with the base because you know its concentration, and you do not know the concentration of the acid. (Note 1000 cm³ = 1 l.)

$$\text{number of moles} = {\text{volume} \atop (l)} \times {\text{concentration} \atop (mol/l)}$$

$$\text{number of moles} \atop \text{of NaOH} = {\text{volume} \atop (25.0 \, cm^3)} \times {\text{concentration} \atop (0.100 \, mol/l)}$$

$$= 25.0 \times 10^{-3} l \times 0.100 \, mol/l = 2.50 \times 10^{-3} \, mol$$

3 Now work out the concentration of the acid.

number of moles of HCl = number of moles of NaOH
$$= 2.50 \times 10^{-3} \, mol$$

$$\text{number of moles} \atop \text{of HCl} = {\text{volume of} \atop HCl(aq)} \times {\text{concentration} \atop \text{of } HCl(aq)}$$

Therefore, if c mol/l is the concentration of HCl,

$$2.50 \times 10^{-3} \, mol = 15.0 \times 10^{-3} l \times c \, mol/l$$

$$c \, mol/l = \frac{2.50 \times 10^{-3} \, mol}{15.0 \times 10 \, l}$$

$$= \mathbf{0.167 \, mol/l}$$

Example 2

In a titration, 25.0 cm³ of sulphuric acid of concentration 0.150 mol/l neutralised 31.2 cm³ of potassium hydroxide solution. Find the concentration of the potassium hydroxide solution.

1 First write the equation:

$$\text{sulphuric} \atop \text{acid} \; + \; \text{potassium} \atop \text{hydroxide} \; \rightarrow \; \text{potassium} \atop \text{sulphate} \; + \; \text{water}$$

$$H_2SO_4(aq) + 2KOH(aq) \rightarrow K_2SO_4(aq) + 2H_2O(l)$$

The equation tells you that 1 mol of H_2SO_4 neutralises 2 mol of KOH.

2 Now work out the number of moles of acid. You must start with the acid because you do not know the concentration of the base.

$$\text{number of moles of acid} \atop (l) = {\text{volume} \atop (25.0 \, cm^3)} \times {\text{concentration} \atop (0.150 \, mol/l)}$$

$$= 25.0 \times 10^{-3} l \times 0.150 \, mol/l = 3.75 \times 10^{-3}$$

3 Now work out the concentration of base.

number of moles of KOH = 2 × number of moles of H_2SO_4
$$= 7.50 \times 10^{-3}$$

$$\text{number of moles} \atop \text{of KOH} = {\text{volume} \atop \text{of KOH(aq)}} \times {\text{concentration} \atop \text{of KOH(aq)}}$$

Therefore, if c mol/l is the concentration of KOH,

$$c \, mol/l = \frac{7.50 \times 10^{-3} \, mol}{31.2 \times 10^{-3} l}$$

$$= \mathbf{0.240 \, mol/l}$$

round-up

on titration

Work out your score on page 114.

1 25.0 cm³ of sodium hydroxide solution are neutralised by 15.0 cm³ of a solution of hydrochloric acid of concentration 0.25 mol/l.
Find the concentration of the sodium hydroxide solution. [1]

2 A solution of sodium hydroxide contains 10 g/l of solute.
  a) What is the concentration of the solution in mol/l? [1]
  b) What volume of this solution would be needed to neutralise 25.0 cm³ of 0.10 mol/l hydrochloric acid? [1]

3 The class decides to test some antacid indigestion tablets. They dissolve the tablets and titrate the alkali in them against a standard acid. Their results are shown in the table.

brand	price of 100 tablets/ £	volume of 0.01 mol/l acid required to neutralise one tablet
Stoppo	0.76	2.8 cm³
Settlo	0.87	3.0 cm³
Alko	1.08	3.3 cm³
Baso	1.30	3.6 cm³

  a) Which antacid tablets offer the best value for money? [1]
  b) What other factors would you consider before choosing a brand? [2]

4 A tanker of acid is emptied into a water supply by mistake. A chemist titrates the water and finds that 10.0 litres of water are needed to neutralise 10.0 cm³ of a 0.010 mol/l solution of sodium hydroxide. What is the concentration of hydrogen ions in the water? [2]

Well done if you've improved. Don't worry if you haven't. Take a break and try again.

preview

At the end of this section you will understand:

- **how metals combine to form chemical cells**
- **oxidation–reduction reactions**
- **redox cells**
- **the electrochemical series.**

How much do you already know? Work out your score on page 114.

Test yourself

1 Refer to the electrochemical series (page 72). Which of these pairs of metals would give the highest reading on the voltmeter when made into a chemical cell?
a) Mg + Cu
b) Mg + Zn
c) Al + Zn
d) Al + Cu
e) Sn + Ni
f) Zn + Pb   [1]

2

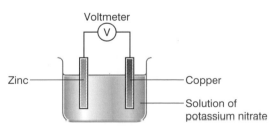

Voltmeter

Zinc — Copper

Solution of potassium nitrate

a) In which direction will electrons travel through the voltmeter?   [1]
b) If copper is replaced by iron, what will happen to the reading on the voltmeter?   [1]

3 a) Complete the following equations. If no reaction happens, write n.r.
(i) zinc(s) + magnesium sulphate(aq) →
(ii) magnesium(s) + zinc sulphate(aq) →
(iii) aluminium(s) + potassium chloride(aq) →
(iv) copper(s) + silver nitrate(aq) →
(v) copper(s) + iron(II) sulphate(aq) →
(vi) iron(s) + copper(II) sulphate(aq) →   [9]
‖ b) Give ionic equations for the reactions in part a).   [12]

4 State which substance is oxidised and which is reduced in each of these reactions.
a) $2Al(s) + Fe_2O_3(s) → 2Fe(s) + Al_2O_3(s)$
b) $Mg(s) + 2H^+(aq) → Mg^{2+}(aq) + H_2(g)$
c) $Hg(l) + I_2(s) → HgI_2(s)$
d) $2SO_2(g) + O_2(g) → 2SO_3(g)$ [8]

10.1 Batteries and cells

An electric current is a stream of electrons. We often use a **battery** as a portable source of electricity. A chemical reaction inside the battery produces an electric current. Lead-acid batteries are used in motor vehicles. The reaction between the lead electrodes and the sulphuric acid electrolyte produces a current. When the chemicals have been used up the battery is recharged: it is connected to a power supply which reverses the chemical reaction and enables the battery to be used again.

A container in which a chemical reaction gives rise to an electric current is called a **chemical cell**. Cells can be connected in a series to form a battery. A 12 volt lead-acid battery consists of six 2 volt cells connected in series. We will now look at the way in which chemical reactions give rise to electric currents.

10.2 Metals in chemical cells

Metals are composed of positive ions and a cloud of electrons. When a reactive metal such as zinc is put into water or a solution of an electrolyte, some zinc

THE METALLIC BOND Page 75.

ions pass into solution. Electrons accumulate on the metal until the negative charge on the zinc becomes high enough to prevent any more Zn^{2+} ions from leaving the metal. Copper is low in the reactivity series and copper ions have little tendency to pass into solution.

THE REACTIVITY SERIES Page 72.

A strip of zinc and a strip of copper may be immersed in a solution and then connected, as shown below. Electrons flow through the external circuit from zinc, which is negatively charged, to copper, which has very little charge.

This kind of cell is called a **chemical cell**. The chemical reaction taking place inside the cell causes a current to flow through the external

circuit. Many metals can be paired up in chemical cells like this. The direction of flow of electrons is from a metal higher in the reactivity series to a metal

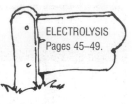

ELECTROLYSIS Pages 45–49.

lower in the series. The wider apart the metals are in the reactivity series, the greater is the voltage of the cell.

Dry cells

Dry cells are chemical cells used in batteries for torches, radios, etc. A dry cell has a damp paste instead of a liquid electrolyte, so it cannot leak.

★ The **zinc–carbon** dry cell has a voltage of 1.5 V.

★ The **alkaline manganese** cell uses zinc and manganese(IV) oxide, and also has a voltage of 1.5 V.

★ **Silver oxide** cells are the tiny 'batteries' used in electronic wristwatches and cameras.

★ The **nickel–cadmium** cell can be recharged when it goes 'flat'. This is done by connecting the cell to a source of direct current. The chemical reaction is reversed, and the cell has a new source of voltage.

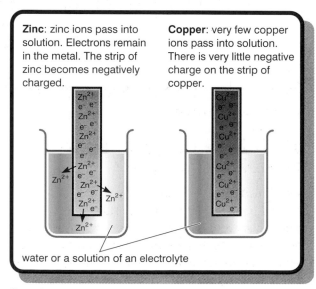

Zinc: zinc ions pass into solution. Electrons remain in the metal. The strip of zinc becomes negatively charged.

Copper: very few copper ions pass into solution. There is very little negative charge on the strip of copper.

water or a solution of an electrolyte

Zinc and copper

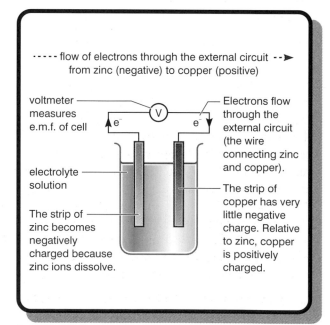

- - - - flow of electrons through the external circuit - - ►
from zinc (negative) to copper (positive)

voltmeter measures e.m.f. of cell

Electrons flow through the external circuit (the wire connecting zinc and copper).

electrolyte solution

The strip of zinc becomes negatively charged because zinc ions dissolve.

The strip of copper has very little negative charge. Relative to zinc, copper is positively charged.

A zinc–copper chemical cell

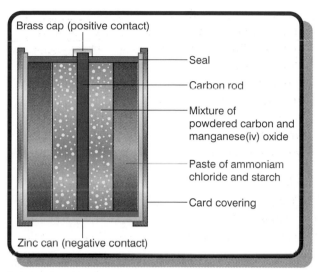

A zinc–carbon dry cell

10.3 Cells

In the zinc–copper cell described in on page 68, the chemical reaction taking place inside the cell causes a current to flow through the external circuit. Chemical energy is converted into electrical energy. This kind of cell is called a **chemical cell**. The voltage obtained from the simple cell shown in the figure on page 68 falls quickly because of changes at the electrodes. For example, the strip of zinc becomes coated with copper due to the displacement reaction:

zinc + copper ions → zinc ions + copper

Changes of this kind which occur at the electrodes are called **polarisation**. In the zinc–copper cell, the polarisation can be stopped by dividing the cell into two half-cells. One way of doing this is shown in the figure opposite. The two half-cells are connected by an **ion bridge** which contains a solution of an electrolyte. A solution of potassium chloride in agar gel is often used. This is made by dissolving agar in a warm solution of potassium chloride, drawing this solution into a length of bent glass tubing and waiting for the solution to cool and the gel to set. The result is a solid gel which conducts and joins the two half-cells.

Ions flow through the ion bridge. In the zinc half-cell, Zn^{2+} ions are formed. To balance the excess of positive charge, some Zn^{2+} ions move into the

ion bridge and some anions, in this case Cl^- ions, move from the ion bridge into the zinc half-cell. In the copper half-cell, Cu^{2+} ions are discharged. To balance the excess of negative charge some SO_4^{2-} ions move into the ion bridge and some cations, e.g. K^+, move from the ion bridge into the copper half-cell. Overall, cations tend to move from the zinc half-cell into the copper half-cell and anions tend to move from the copper half-cell into the zinc half-cell. Neither half-cell has a net positive or negative charge at any time.

The movement of ions across the ion bridge is very slow and extensive mixing does not occur. The cell reaction cannot take place without a ion bridge. Nor can it take place if the half-cells are connected by an electrical conductor such as copper wire.

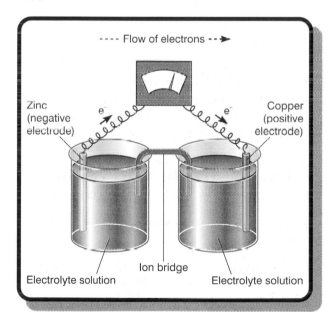

A chemical cell with an ion bridge

10.4 How simple can a chemical cell get?

Metals differ in the readiness with which they part with electrons. The more reactive metals are more ready to part with electrons than the less reactive metals. What happens when different metals are connected by a conducting circuit? You can find out by making a cell from a lemon; as shown on page 70.

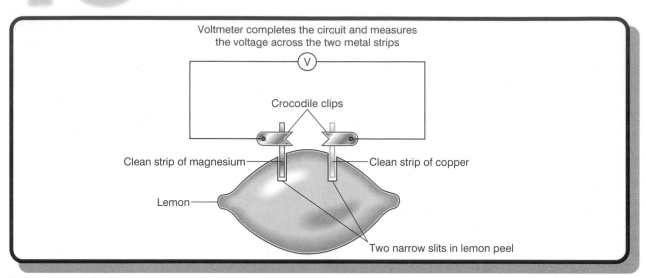

Voltmeter completes the circuit and measures the voltage across the two metal strips

Crocodile clips

Clean strip of magnesium

Clean strip of copper

Lemon

Two narrow slits in lemon peel

Making a cell from a lemon

Measurements with pairs of metals used to make a lemon cell give voltages in the order:

Mg-Cu > Zn-Cu > Fe-Cu > Al-Cu > Pb-Cu

You may not always have a lemon handy when you want to make a chemical cell. If so, try dipping strips of two different metals into a solution of sodium chloride (1 mol/l). Connect the metals to a voltmeter.

10.5 Oxidising and reducing agents

Oxidation–reduction reactions

Do you remember what is meant by an oxidising agent and a reducing agent?

★ An **oxidising agent** gives oxygen to a substance or takes hydrogen from a substance.

★ A **reducing agent** takes oxygen from a substance or gives hydrogen to a substance. For example, hydrogen reduces lead(II) oxide to lead.

```
        ┌─── this is reduction ───┐
lead(II) oxide + hydrogen ──────→ lead + water
      PbO(s) + H₂(g)  ──────→ Pb(s) + H₂O(l)
              └─── this is oxidation ───┘
```

$$PbO(s) + H_2(g) \longrightarrow Pb(s) + H_2O(l)$$

In this reaction, lead(II) oxide is reduced while hydrogen is oxidised. Lead(II) oxide is the

oxidising agent, and hydrogen is the reducing agent. Oxidation and reduction always occur together, and these reactions are called **oxidation–reduction reactions** or **redox reactions**.

Metals as reducing agents

Metals are reducing agents; that is, they can accept oxygen from other substances, for example:

copper + oxygen → copper(II) oxide

$$2Cu(s) + O_2(g) \longrightarrow 2CuO(s)$$

Note that when copper acts as a reducing agent copper is oxidised, and Cu atoms lose electrons to form Cu^{2+} ions. Oxidation involves the loss of electrons.

calcium + water → hydrogen + calcium hydroxide

$$Ca(s) + 2H_2O(l) \rightarrow H_2(g) + Ca(OH)_2(aq)$$

Note that calcium atoms have lost electrons to form Ca^{2+} ions. Oxidation involves the loss of electrons.

Some metals are more powerful reducing agents than others. The higher a metal is in the reactivity series, the more powerful a reducing agent it is. A metal

REACTIVITY SERIES Page 72.

high in the reactivity series can reduce the oxide of a metal lower in the reactivity series, for example:

$$\text{aluminium} + \underset{\text{oxide}}{\text{iron(III)}} \rightarrow \underset{\text{oxide}}{\text{aluminium}} + \text{iron}$$

$$2Al(s) + Fe_2O_3(s) \rightarrow Al_2O_3(s) + 2Fe(s)$$

This is an oxidation–reduction reaction: while iron(III) oxide is reduced to iron, aluminium is oxidised to aluminium oxide.

The halogens as oxidising agents

The halogens are classified as oxidising agents, although they do not give oxygen to other substances. Their reactions are dominated by a readiness to gain electrons and form halide ions, for example:

HALOGENS
Page 23.

$$\underset{\text{molecules}}{\text{chlorine}} + \text{electrons} \rightarrow \text{chloride ions}$$

$$Cl_2(aq) + 2e^- \rightarrow 2Cl^-(aq)$$

To include halogens in the definition of oxidising agents, our definition of oxidation–reduction (redox) reactions must be extended.

An oxidising agent is defined as a substance that can take electrons from another substance. The oxidising agent is reduced in the reaction. The substance that loses electrons is oxidised.

An extended definition:

Oxidation is	**Reduction is**
• the gain of oxygen by a substance	• the loss of oxygen by a substance
• the loss of hydrogen by a substance	• the gain of hydrogen by a substance
• the loss of electrons by a substance.	• the gain of electrons by a substance.

An oxidising agent	**A reducing agent**
• gives oxygen to a substance	• takes oxygen from a substance
• takes hydrogen from a substance	• gives hydrogen to a substance
• takes electrons from a substance.	• gives electrons to a substance.

Handy hint

Loss or gain of electrons?

OIL RIG will help you to remember:

★ Oxidation Is Loss

★ Reduction Is Gain.

Reactions with metals: halogens react with metals to form salts, for example:

$$\text{sodium} + \text{chlorine} \rightarrow \text{sodium chloride}$$

$$2Na(s) + Cl_2(g) \rightarrow 2Na^+(s) + 2Cl^-(s)$$

Halogen molecules are reduced (they gain electrons) to become halide ions, such as Cl^-.

Metal atoms are oxidised (they give electrons) to become metal ions, such as Na^+.

10.6 Displacement reactions

When a strip of zinc is added to a solution of copper(II) sulphate solution, some of the zinc goes into solution as zinc ions. The blue colour of the solution fades and a reddish brown deposit of solid copper forms. Zinc has *displaced* copper from copper(II) sulphate.

$$Zn(s) + CuSO_4(aq) \rightarrow Cu(s) + ZnSO_4(aq)$$

The reaction is a **displacement reaction**. Zinc is a more powerful reducing agent than copper, therefore, the reactions that happen are:

$$Zn(s) \rightarrow Zn^{2+}(aq) + 2e^- \quad \text{(This is oxidation)}$$

$$Cu^{2+}(aq) + 2e^- \rightarrow Cu(s) \quad \text{(This is reduction)}$$

These are the same reactions as are happening in the cell shown on page 68. The overall cell reaction is:

$$Zn(s) + Cu^{2+}(aq) \rightarrow Zn^{2+}(aq) + Cu(s)$$

This is an oxidation–reduction reaction, a redox reaction.

Metals high in the electrochemical series are more powerful reducing agents than metals lower in the series (see page 72). A metal high in the electrochemical series will therefore displace a metal below it in the reactivity series from its salts.

10.7 The electrochemical series

The electrochemical series puts all oxidising agents and reducing agents in order. It includes metals, which are reducing agents and oxidising agents such as the halogens and potassium permanganate.

A section of the electrochemical series

Potassium   Most powerful reducing agents

Calcium

Sodium

Magnesium

Aluminium

Zinc

Iron

Tin

Lead

Hydrogen

Copper

Iron(III) ions, Fe^{3+}

Silver

Mercury

Bromine Br_2

Potassium dichromate, acid $K_2Cr_2O_7$, H^+

Chlorine, Cl_2

Potassium permanganate, acid $KMnO_4$, H^+

Fluorine, F_2

Most powerful oxidising agents

The electrochemical series does not always show the true ability of a metal to act as a reducing agent. Calcium appears to be less reactive than sodium. This is because the reactions of calcium are slowed down by the formation of a layer of insoluble calcium compound on the surface of the metal. Aluminium appears less reactive than the electrochemical series shows because aluminium is coated with a protective layer of aluminium oxide.

10.8 Redox cells with ion bridges

The cell shown in the figure on page 69 connects two metal–metal ion systems with an ion bridge. The current which flows through the external circuit depends on the difference between the positions of the two metals in the electrochemical series. Any redox system involved the exchange of electrons:

Reducing agent $\rightleftharpoons$ Oxidising agent + electrons

When two redox systems are connected by an ion bridge, a current will flow through the external circuit, as shown in the diagram below.

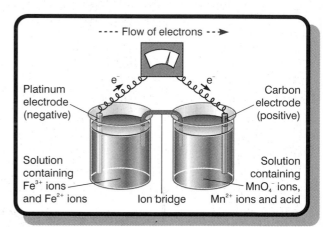

A redox cell with an ion bridge

Oxidation

$$Fe^{2+}(aq) \rightarrow Fe^{3+}(aq) + e^-$$

The electrode gains electrons and becomes negatively charged. The solution changes colour from green to reddish brown.

Reduction

$$MnO_4{}^-(aq) + 8H^+(aq) + 5e^- \rightarrow$$
$$Mn^{2+}(aq) + 4H_2O(l)$$

The electrode loses electrons and becomes positively charged. The solution changes colour from purple to pale pink.

round-up

How much have you improved?
Work out your improvement index on page 114–115.

1 The diagram shows a cell made from everyday materials.

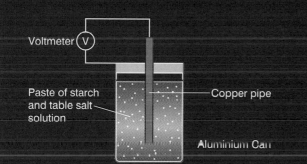

Voltmeter (V)

Paste of starch and table salt solution

Copper pipe

Aluminium Can

a) Show the direction in which electrons pass through the wire. [1]
b) Why could the paste not be made from starch and sugar solution? [1]
c) Would the reading on the voltmeter be higher or lower if a tin can were used? [1]

2

Ammeter
(mA)

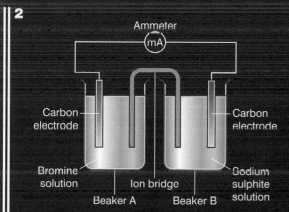

Carbon electrode

Carbon electrode

Bromine solution

Ion bridge

Sodium sulphite solution

Beaker A   Beaker B

The reaction that takes place in Beaker B is the oxidation:

$$SO_3^{2-}(aq) + H_2O(l) \rightarrow SO_4^{2-}(aq) + 2H^+(aq) + 2e^-$$

a) Write the ion–electron equation for the reduction of bromine in Beaker A. [2]
b) In which direction do electrons flow through the ammeter? [1]
c) What is the purpose of the ion bridge? [1]

3 Refer to the lemon cell on page 70.
a) Which metal is the positive in each cell? [1]
b) How does the order of voltages help you to classify the metals? [1]
c) In which direction do electrons flow through the external circuit? [1]
d) Name a metal which would form a couple with copper in which the electron flow was in the reverse direction. [1]

4 Refer to the iron(II)–permanganate cell on page 72.
a) In which direction do electrons flow through the voltmeter? [1]
b) Which electrode process is oxidation? [1]
c) If the beaker on the right contains potassium iodide solution and the beaker on the left contains iron(II) and iron(III) sulphate solution as before, write ion–electron equations for the electrode processes involving **(i)** I⁻ and **(ii)** Fe^{3+}. [4]
d) Combine these equations to give the equation for the overall cell reaction. [4]
e) In which direction do electrons flow through the voltmeter? [1]

Well done if you've improved. Don't worry if you haven't. Take a break and try again.

Metals and alloys

preview

At the end of this topic you will:

- understand the nature of the metallic bond
- be familiar with the chemical reactions of metals
- use the reactivity series to make predictions about metals and their compounds and methods used for extracting metals
- calculate the empirical formula and molecular formula of a compound
- use the equation for a reaction to calculate the masses of solids that react.

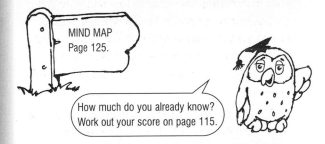

MIND MAP
Page 125.

How much do you already know?
Work out your score on page 115.

Test yourself

1 List three characteristics of metals that are explained by the metallic bond. [3]

2 Name three metals that burn in air to form oxides. [3]

3 Name two metals that do not react when heated in air. [2]

4 Name three metals that react with cold water and say what products are formed. [5]

5 What is formed when a metal reacts with hydrochloric acid? [2]

6 In which groups of the periodic table do you find
a) sodium **b)** magnesium **c)** transition metals? [3]

7 What method is used to extract very reactive metals from their ores? [2]

8 How is iron extracted from its ore? [3]

9 Name two metals which become coated with a film of oxide on exposure to the air. [2]

10 State which of the following oxides is/are:
a) soluble in water
b) decomposed by heat alone
c) reduced by heating with carbon
d) amphoteric.
Na_2O, Al_2O_3, ZnO, CuO, PbO, Ag_2O [7]

11 A Silver
B Carbon
C Zinc
D Chlorine
E Argon
F Iodine.
Identify **a)** the metal used for galvanising iron **b)** the metal found uncombined in the Earth's crust **c)** the very unreactive element **d)** two non-metallic elements with similar chemical properties. [5]

12 A calcium
B aluminium
C potassium
D lead
E iron.
Which metal or metals in this list can be described as **a)** a transition metal **b)** a catalyst for the manufacture of ammonia **c)** having the highest density **d)** the most reactive **e)** reacting with water to give an alkaline solution? [6]

13 Calculate the empirical formulae of
a) the compound containing 55.5% mercury and 44.5% bromine by mass [1]
b) the compound formed from 14.9 g of copper and 17.7 g of chlorine [1]
c) the compound formed when 0.69 g of sodium forms 0.93 g of an oxide. [1]

14 Calculate the percentage by mass of sulphur in
a) SO_2 **b)** SO_3 **c)** H_2SO_4. [3]

15 When a mixture of 8 g of iron and 4 g of sulphur is heated, the elements react to form iron(II) sulphide, FeS. How much iron will be left over at the end of the reaction? [1]

11.1 The metallic bond

A block of metal consists of positive metal ions and free electrons, as shown in the diagram.

The presence of free electrons explains how metals can conduct electricity. Electrons can be supplied at one end of a piece of metal and removed at the other end. The nature of the metallic bond also explains how metals can change their shape without breaking.

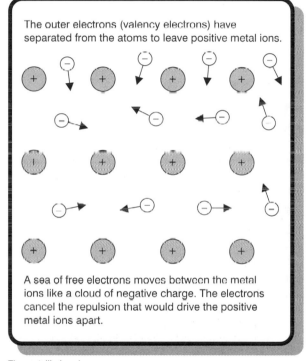

The outer electrons (valency electrons) have separated from the atoms to leave positive metal ions.

A sea of free electrons moves between the metal ions like a cloud of negative charge. The electrons cancel the repulsion that would drive the positive metal ions apart.

The metallic bond

11.2 Why are different metals used for different purposes?

A metal is chosen for a particular purpose on the basis of its characteristics.

★ **Melting point:** Mercury is a liquid (m.p. –39 °C). Mixtures of mercury with other metals are called **amalgams**. They are solids with low melting points and are used in dentistry to fill cavities because they can be moulded at the temperature of the mouth before they set hard.

★ **Appearance:** Gold is used in jewellery because of its beautiful colour and shine, and because it does not corrode.

★ **Electrical conductivity:** Copper is used in electrical wiring because of its high electrical conductivity and freedom from corrosion. Other metals also have good electrical conductivity.

★ **Cost:** The cost of metals comes into the choice. Gold wire is not used for wiring a house although gold never corrodes. Copper is not completely free from corrosion, but it is much less costly than gold. In space capsules, resistance to corrosion is of prime importance and no expense is spared, so gold is used for electrical contacts.

★ **Thermal conductivity:** Copper, aluminium and iron are used for saucepans because of their high thermal conductivity.

★ **Strength:** Iron is used for the manufacture of machines, bridges and ships because of its strength.

★ **Density:** Aluminium alloys are used for the manufacture of doors, window frames and small boats because of their low density and freedom from corrosion. Lead is used as protection from radiation because of its high density.

★ **Malleability** (the ability to be hammered into different shapes): Cast iron is treated to remove carbon and make it malleable for use in the car industry as wheel hubs, brake shoes, door hinges etc.

11.3 Reactions of metals

Most metals react slowly with air to form a surface film of metal oxide. This reaction is called **tarnishing**. Gold and platinum do not tarnish in air. Aluminium rapidly forms a surface layer of aluminium oxide and only shows its true reactivity if this layer is removed. The table on the following page shows some of the reactions of metals. Equations are given for some of them.

metal	reaction when heated in oxygen	reaction with cold water	reaction with dilute hydrochloric acid
potassium sodium lithium calcium	burn to form the oxides	displace hydrogen; form alkaline hydroxides.	react dangerously fast to form hydrogen and the metal chloride
magnesium aluminium zinc iron		reacts slowly do not react, except for slow rusting of iron; all react with steam	displace hydrogen; form metal chlorides
tin lead copper	slowly form oxides without burning	do not react even with steam	react very slowly to form hydrogen and the metal chloride
silver gold platinum	do not react		do not react

sodium + water → hydrogen + sodium hydroxide

$$2Na(s) + 2H_2O(l) \rightarrow H_2(g) + 2NaOH(aq)$$

(Sodium must be kept under oil to prevent it being attacked by water in the air.)

calcium + water → hydrogen + calcium hydroxide

$$Ca(s) + 2H_2O(l) \rightarrow H_2(g) + Ca(OH)_2(aq)$$

magnesium + water → hydrogen + magnesium hydroxide

$$Mg(s) + 2H_2O(l) \rightarrow H_2(g) + Mg(OH)_2(aq)$$

zinc + hydrochloric acid → hydrogen + zinc chloride

$$Zn(s) + 2HCl(aq) \rightarrow H_2(g) + ZnCl_2(aq)$$

The metals can be placed in an order of reactivity which is called the **reactivity series**:

potassium	K	
sodium	Na	
lithium	Li	increase in reactivity ↑
calcium	Ca	
magnesium	Mg	
aluminium	Al	increase in
zinc	Zn	the ease
iron	Fe	with which
tin	Sn	metals
lead	Pb	react to
copper	Cu	form ions
silver	Ag	
gold	Au	
platinum	Pt	

Part of the reactivity series of metals

11.4 Metals in the periodic table

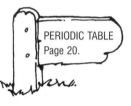
PERIODIC TABLE
Page 20.

In Group 1 of the periodic table are the **alkali metals**, and in Group 2 are the **alkaline earths**. Aluminium is in Group 3. The less reactive metals tin and lead are in Group 4. The metals in the block between Group 2 and Group 3 are called the **transition metals**, e.g. iron, nickel, copper and zinc. For the differences between the physical and chemical properties of metallic and non-metallic elements, see page 21.

11.5 Making predictions

Competition between metals to form ions

Metals high in the reactivity series form ions with ease. A metal which is higher in the reactivity series will displace a metal which is lower in the reactivity series from a salt. Examples are:

copper + silver nitrate solution (colourless solution) → silver (silver crystals, turn black in light) + copper(II) nitrate solution (blue solution)

$$Cu(s) + 2AgNO_3(aq) \rightarrow 2Ag(s) + Cu(NO_3)_2(aq)$$

or $$Cu(s) + 2Ag^+(aq) \rightarrow Cu^{2+}(aq) + 2Ag(s)$$

zinc + copper(II) sulphate → copper + zinc sulphate

(blue solution)   (reddish   (colourless
                     brown solid) solution)

$$Zn(s) + CuSO_4(aq) \rightarrow Cu(s) + ZnSO_4(aq)$$

or $$Zn(s) + Cu^{2+}(aq) \rightarrow Zn^{2+}(aq) + Cu(s)$$

Compounds and the reactivity series

Hints & Tips

The higher a metal is in the reactivity series,

*   the more readily it forms compounds
*   the more difficult it is to split up its compounds.

Oxides

★ Hydrogen will reduce the oxides of metals which are low in the reactivity series, e.g.

copper(II) oxide + hydrogen $\xrightarrow{\text{heat}}$ copper + water

$$CuO(s) + H_2(g) \rightarrow Cu(s) + H_2O(l)$$

★ Carbon, when heated, will reduce the oxides of metals which are low in the reactivity series, e.g.

lead(II) oxide + carbon $\xrightarrow{\text{heat}}$ lead + carbon monoxide

$$PbO(s) + C(s) \rightarrow Pb(s) + CO(g)$$

★ Carbon monoxide is used to reduce hot iron oxide to iron.

iron(III) oxide + carbon monoxide $\xrightarrow{\text{heat}}$ iron + carbon dioxide

$$Fe_2O_3(s) + 3CO(g) \rightarrow 2Fe(s) + 3CO_2(g)$$

★ The oxides of metals which are high in the reactivity series, e.g. aluminium, are not reduced by hydrogen or carbon or carbon monoxide.

★ Silver and mercury are very low in the reactivity series. Their oxides decompose when heated.

11.6 Extracting metals

The method chosen for extracting a metal from its ore depends on the position of the metal in the reactivity series; see table opposite.

Metals which are very low in the reactivity series occur 'native', that is as the free metal. Copper,

silver and gold have been known to mankind for thousands of years because they occur native. Metals which are low in the reactivity series, e.g. tin and lead, are relatively easy to obtain from their ores by reduction. These metals have been known for longer than highly reactive metals such as aluminium and titanium. The metals at the top of the reactivity series are obtained by electrolysis and have been obtained only in the last two centuries.

Methods of extracting metals from their ores are reduction reactions, e.g. reduction of iron(III) oxide in the blast furnace:

iron(III) oxide + carbon monoxide → iron + carbon dioxide

$$Fe_2O_3(s) + 3CO(g) \rightarrow 2Fe(s) + 3CO_2(g)$$

e.g. electrolysis of molten aluminium oxide:

aluminium ions → aluminium atoms

$$Al^{3+}(l) + 3e^- \rightarrow Al(l)$$

Iron

The chief ores of iron are haematite, Fe_2O_3, magnetite, Fe_3O_4, and iron pyrites, FeS_2. The sulphide ore is roasted in air to convert it into an oxide. The oxide ores are reduced to iron in a blast furnace (see diagram below). The blast furnace is run continuously. The low cost of extraction and the plentiful raw materials make iron cheaper than other metals.

potassium sodium calcium magnesium	}	Anhydrous chloride is melted and electrolysed.
aluminium	—	Molten anhydrous oxide is electrolysed.
zinc iron lead	}	Sulphides are roasted to give oxides which are reduced with carbon; oxides are reduced with carbon.
copper	—	Sulphide ore is heated with a controlled volume of air.
silver gold	}	Found 'native' (as the free metals).

FACTS

Methods used for the extraction of metals from their ores

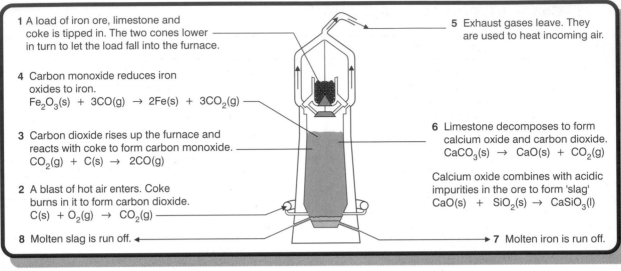

1 A load of iron ore, limestone and coke is tipped in. The two cones lower in turn to let the load fall into the furnace.

4 Carbon monoxide reduces iron oxides to iron.
$Fe_2O_3(s) + 3CO(g) \rightarrow 2Fe(s) + 3CO_2(g)$

3 Carbon dioxide rises up the furnace and reacts with coke to form carbon monoxide.
$CO_2(g) + C(s) \rightarrow 2CO(g)$

2 A blast of hot air enters. Coke burns in it to form carbon dioxide.
$C(s) + O_2(g) \rightarrow CO_2(g)$

8 Molten slag is run off.

5 Exhaust gases leave. They are used to heat incoming air.

6 Limestone decomposes to form calcium oxide and carbon dioxide.
$CaCO_3(s) \rightarrow CaO(s) + CO_2(g)$

Calcium oxide combines with acidic impurities in the ore to form 'slag'
$CaO(s) + SiO_2(s) \rightarrow CaSiO_3(l)$

7 Molten iron is run off.

A blast furnace

11.7 Iron and steel

Cast iron

The iron that comes out of the blast furnace is called cast iron. It contains 3–4% carbon which lowers the melting point, making cast iron easier to melt and mould than pure iron. By casting, objects with complicated shapes can be made, such as the engine blocks of motor vehicles. The carbon content makes cast iron brittle.

Steel

Steel is made from iron by burning off carbon and other impurities in a stream of oxygen. A number of elements may be added to give different types of steel:

- chromium in stainless steels for cutlery, car accessories and tools
- cobalt steel in permanent magnets
- manganese in all steels to increase strength
- molybdenum steel for rifle barrels and propeller shafts
- nickel in stainless steel cutlery and industrial plants
- tungsten steel in high-speed cutting tools
- vanadium steel in springs.

11.8 Conservation

The Earth's resources of metals are limited. It makes sense to collect scrap metals and recycle them. In addition, there is a saving in fuel resources because less energy is needed for recycling than for extracting metals from their ores. There is another reason for conserving metals: the impact which mining has on the environment. Before recycling, scrap metals must be collected, sorted and stored until there is enough to process.

11.9 Uses of metals and alloys

metal/alloy	characteristics	uses
aluminium	low density good electrical conductor good thermal conductor reflector of light non-toxic resistant to corrosion	aircraft manufacture (Duralumin) overhead electrical cable saucepans, etc. car headlamps food packaging door frames, window frames, etc.
brass, an alloy of copper and zinc	golden colour, harder than copper, resists corrosion	ships' propellers, taps, screws, electrical fittings
bronze, an alloy of copper and tin	golden colour, hard, sonorous, resistant to corrosion	coins, medals, statues, springs, church bells
copper	good electrical conductor not corroded readily	electrical circuits water pipes and tanks
Duralumin, an alloy of aluminium	low density, stronger than aluminium	aircraft and spacecraft
gold	beautiful colour never tarnishes	jewellery, dentistry, electrical contacts
iron	hard, strong, inexpensive, rusts	construction, transport
lead	dense, unreactive, soft, not very strong	car batteries, divers' weights, roofing
magnesium	bright flame	flares and flash bulbs
mercury	liquid at room temperature	thermometers, dental amalgam for filling teeth
nickel	resists corrosion, strong, tough, hard	stainless steel
silver	beautiful colour and shine good electrical conductor good reflector of light	jewellery, silverware contacts in computers, etc. mirrors, dental amalgam
solder, alloy of tin and lead	low melting point	joining metals in an electrical circuit
steel, an iron alloy	strong	buildings, machinery, transport
tin	low in reactivity series	coating 'tin cans'
titanium	low in density, strong, very resistant to corrosion	high-altitude planes, nose-cones of spacecraft
zinc	high in reactivity series	protection of iron and steel by galvanising

The uses of some metals and alloys

11.10 Formulae

Molecular formula

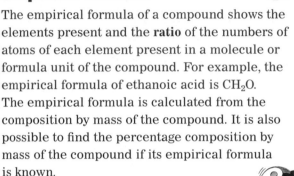

The molecular formula of a compound shows the elements present and how many atoms of each element are present in a molecule or formula unit of the compound. For example, the molecular formula of ethanoic acid is $C_2H_4O_2$.

Empirical formula

The empirical formula of a compound shows the elements present and the **ratio** of the numbers of atoms of each element present in a molecule or formula unit of the compound. For example, the empirical formula of ethanoic acid is CH_2O. The empirical formula is calculated from the composition by mass of the compound. It is also possible to find the percentage composition by mass of the compound if its empirical formula is known.

Calculation of empirical formula

3.72 g of phosphorus react with 4.80 g of oxygen to form an oxide. What is the empirical formula of the oxide?

	phosphorus		oxygen
mass	3.72 g		4.80 g
RAM	31		16
amount (moles)	$\frac{3.72}{31}$		$\frac{4.80}{16}$
	= 0.12		0.30
ratio of moles	= 1	to	2.5
ratio of numbers of atoms	= 2	to	5

Empirical formula is P_2O_5.

Calculation of percentage composition

Find the percentages by mass of carbon, hydrogen and oxygen in propanol, C_3H_7OH.

The empirical formula is C_3H_8O.

$RFM = (3 \times 12) + 8 + 16 = 60$

$$\text{percentage of carbon} = \frac{RAM(C) \times \text{no. of C atoms}}{RFM(C_3H_8O)} \times 100\%$$

$$= \frac{(12 \times 3)}{60} \times 100\% = 60.0\%$$

$$\text{percentage of hydrogen} = \frac{RAM(H) \times \text{no. of H atoms}}{RFM(C_3H_8O)} \times 100\%$$

$$= \frac{(1 \times 8)}{60} \times 100\% = 13.3\%$$

$$\text{percentage of oxygen} = \frac{RAM(O) \times \text{no. of O atoms}}{RFM(C_3H_8O)} \times 100\%$$

$$= \frac{(16 \times 1)}{60} \times 100\% = 26.7\%$$

You can check on your calculation by adding up the percentages:

carbon 60.0% + hydrogen 13.3% + oxygen 26.7% = 100%

Calculation of molecular formula

The molecular formula is a multiple of the empirical formula:

molecular formula = (empirical formula)$_n$

Therefore relative formula mass = $n \times$ relative empirical formula mass.

For ethanoic acid:

empirical formula = CH_2O

relative empirical formula mass = $(12 + 2 + 16) = 30$

The relative formula mass of ethanoic acid is found by experiment to be 60, therefore

molecular formula = $2 \times$ empirical formula

$$= C_2H_4O_2.$$

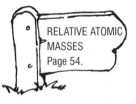

RELATIVE ATOMIC MASSES
Page 54.

round-up

on formulae

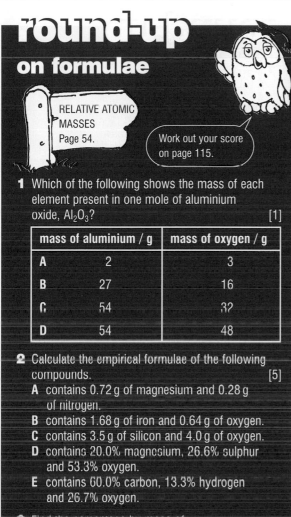

RELATIVE ATOMIC MASSES Page 54.

Work out your score on page 115.

1 Which of the following shows the mass of each element present in one mole of aluminium oxide, Al_2O_3? [1]

	mass of aluminium / g	mass of oxygen / g
A	2	3
B	27	16
C	54	32
D	54	48

2 Calculate the empirical formulae of the following compounds. [5]
 A contains 0.72 g of magnesium and 0.28 g of nitrogen.
 B contains 1.68 g of iron and 0.64 g of oxygen.
 C contains 3.5 g of silicon and 4.0 g of oxygen.
 D contains 20.0% magnesium, 26.6% sulphur and 53.3% oxygen.
 E contains 60.0% carbon, 13.3% hydrogen and 26.7% oxygen.

3 Find the percentage by mass of
 a) carbon and hydrogen in ethane, C_2H_6 [2]
 b) sulphur and oxygen in sulphur trioxide, SO_3 [2]
 c) nitrogen, hydrogen and oxygen in ammonium nitrate, NH_4NO_3 [3]
 d) calcium and bromine in calcium bromide, $CaBr_2$. [2]

• •

TAKE A BREAK

• •

11.11 Masses of reacting solids

The equation for a chemical reaction enables us to calculate the masses of solids that react together, and the masses of solid products that are formed.

Example 1

What mass of copper(II) oxide is formed by the complete oxidation of 3.175 g of copper?

Equation: $2Cu(s) + O_2(g) \rightarrow 2CuO(s)$

The equation shows that 2 mol of copper form 2 mol of copper(II) oxide; that is, 1 mol of copper forms 1 mol of copper(II) oxide.

Putting in the values RAM(Cu) = 63.5, RAM(O) = 16, RFM(CuO) = 79.5:

63.5 g of copper form 79.5 g of copper(II) oxide

3.175 g of copper gives $(\frac{3.175}{63.5}) \times 79.5 = 3.975$ g of copper(II) oxide

Example 2

What mass of calcium carbonate must be decomposed to give 50 tonnes of calcium oxide? (1 tonne = 1×10^3 kg = 1×10^6 g)

Equation: $CaCO_3(s) \rightarrow CaO(s) + CO_2(g)$

Putting in the values RAM(Ca) = 40, RAM(C) = 12, RAM(O) = 16:

RFM of $CaCO_3$ = 40 + 12 + (3 × 16) = 100

RFM of CaO = 40 + 16 = 56

Therefore 56 g of CaO are formed from 100 g of $CaCO_3$.

50 tonnes of CaO are formed from $(50 \times 10^6/56) \times 100$ g of $CaCO_3$ = 89 tonnes of $CaCO_3$

round-up

on reacting masses

RELATIVE ATOMIC MASSES
Page 54.

Work out your score on page 115.

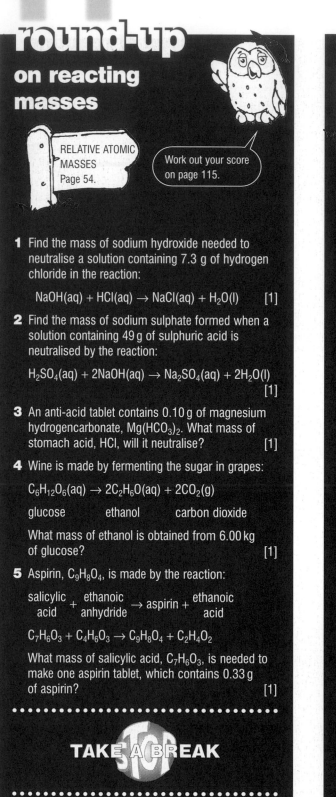

1 Find the mass of sodium hydroxide needed to neutralise a solution containing 7.3 g of hydrogen chloride in the reaction:

$NaOH(aq) + HCl(aq) \rightarrow NaCl(aq) + H_2O(l)$ [1]

2 Find the mass of sodium sulphate formed when a solution containing 49 g of sulphuric acid is neutralised by the reaction:

$H_2SO_4(aq) + 2NaOH(aq) \rightarrow Na_2SO_4(aq) + 2H_2O(l)$ [1]

3 An anti-acid tablet contains 0.10 g of magnesium hydrogencarbonate, $Mg(HCO_3)_2$. What mass of stomach acid, HCl, will it neutralise?   [1]

4 Wine is made by fermenting the sugar in grapes:

$C_6H_{12}O_6(aq) \rightarrow 2C_2H_6O(aq) + 2CO_2(g)$

glucose   ethanol   carbon dioxide

What mass of ethanol is obtained from 6.00 kg of glucose?   [1]

5 Aspirin, $C_9H_8O_4$, is made by the reaction:

$\begin{array}{ccc} \text{salicylic} \\ \text{acid} \end{array} + \begin{array}{c} \text{ethanoic} \\ \text{anhydride} \end{array} \rightarrow \text{aspirin} + \begin{array}{c} \text{ethanoic} \\ \text{acid} \end{array}$

$C_7H_6O_3 + C_4H_6O_3 \rightarrow C_9H_8O_4 + C_2H_4O_2$

What mass of salicylic acid, $C_7H_6O_3$, is needed to make one aspirin tablet, which contains 0.33 g of aspirin?   [1]

TAKE A BREAK

round-up

PERIODIC TABLE
Page 120.

How much have you improved?
Work out your improvement index on pages 115–116.

1 Write equations for the reactions with oxygen of
   a) sodium (to form Na_2O)   [3]
   b) magnesium   [3]
   c) zinc   [3]
   d) iron (to form Fe_3O_4)   [3]
   e) tin (to form SnO)   [3]
   f) lead (to form PbO)   [3]
   g) copper (to form CuO).   [3]

2 a) Write equations for the reactions of
      (i) magnesium and hydrochloric acid   [4]
      (ii) iron and hydrochloric acid (to form $FeCl_2$)   [4]
      (iii) tin and hydrochloric acid (to form $SnCl_2$).   [4]
   b) Write equations for the reactions between
      (i) magnesium and sulphuric acid   [4]
      (ii) iron and sulphuric acid (to give $FeSO_4$).   [4]

3 Copy and complete these word equations. If there is no reaction, write 'no reaction'.   [8]
   a) magnesium + sulphuric acid →
   b) platinum + sulphuric acid →
   c) silver + hydrochloric acid →
   d) gold + hydrochloric acid →
   e) zinc + sulphuric acid →
   f) tin + water →

4 Why are copper and its alloys used as coinage metals in preference to iron?   [3]

5 The following metals are listed in order of reactivity:

   calcium > magnesium > iron > copper

   Describe how the metals follow this order in their reactions with
   a) water   [4]
   b) dilute hydrochloric acid.   [4]

6 What would you see if you dropped a piece of zinc into a test tube of
   a) copper(II) sulphate solution
   b) lead(II) nitrate solution?
   Write word equations and chemical equations for the reactions.   [11]

7 A metal X displaces another metal Y from a solution of a salt of Y. X is displaced by a metal Z from a solution of a salt of X. List the metals in order of reactivity with the most reactive first. [2]

8 The following metals are listed in order of reactivity, with the most reactive first:
Na Mg Al Zn Fe Pb Cu Hg Au
List the metals which
a) occur as the free elements in the Earth's crust [1]
b) react at an observable speed with cold water [2]
c) react with steam but not with cold water [2]
d) react at an observable speed with dilute acids [4]
e) react dangerously fast with dilute acids [1]
f) displace lead from lead(II) nitrate solution. [4]

9 Suggest what method could be used to extract each of the metals A, B, C and D from a chloride ore or an oxide ore.
• Metal A reacts with cold water. [3]
• Metal B reacts only very, very slowly with water. [2]
• Metal C does not react with steam or with dilute hydrochloric acid. [2]
• Metal D when exposed to air immediately becomes coated with a layer of oxide. [3]

10 Predict the reaction of a) rubidium and cold water
b) palladium and dilute hydrochloric acid. [5]

11 List four different uses for aluminium. Say what property of aluminium makes it suitable for each use. [8]

12 List three savings which are made when metal objects are recycled. [3]

13 a) Calculate the mass of carbon dioxide formed by the action of acid on 15 g of calcium carbonate in the reaction [1]

$$CaCO_3(s) + 2HCl(aq) \rightarrow CO_2(g) + CaCl_2(aq) + H_2O(l)$$

14 A sulphuric acid plant uses 2500 tonnes of sulphur dioxide each day. What mass of sulphur must be burned to produce this quantity of sulphur dioxide? [1]

15 a) Why are the alkali metals so-called? [1]
b) For the reaction between potassium and water write (i) a word equation [4]
(ii) a balanced chemical equation [4]
c) How could you test the products of the reaction? [2]

16 Zinc is obtained by heating zinc oxide with carbon. Aluminium is obtained by the electrolysis of molten aluminium oxide. Explain why different methods are chosen for the two metals. [3]

17

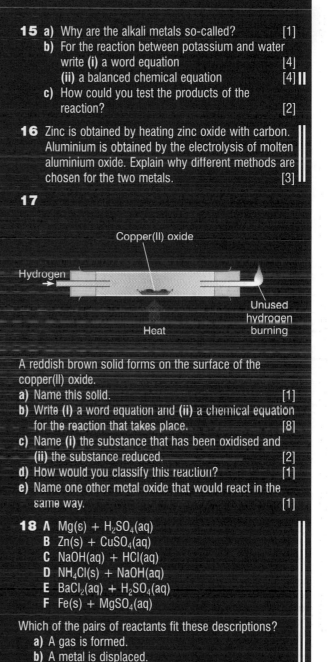

Copper(II) oxide

Hydrogen →

Heat

Unused hydrogen burning

A reddish brown solid forms on the surface of the copper(II) oxide.
a) Name this solid. [1]
b) Write (i) a word equation and (ii) a chemical equation for the reaction that takes place. [8]
c) Name (i) the substance that has been oxidised and (ii) the substance reduced. [2]
d) How would you classify this reaction? [1]
e) Name one other metal oxide that would react in the same way. [1]

18 A $Mg(s) + H_2SO_4(aq)$
B $Zn(s) + CuSO_4(aq)$
C $NaOH(aq) + HCl(aq)$
D $NH_4Cl(s) + NaOH(aq)$
E $BaCl_2(aq) + H_2SO_4(aq)$
F $Fe(s) + MgSO_4(aq)$

Which of the pairs of reactants fit these descriptions?
a) A gas is formed.
b) A metal is displaced.
c) A white precipitate is formed.
d) There is no visible reaction.
e) No reaction takes place. [6]

Corrosion

12

preview

At the end of this topic you will:

- **understand the corrosion of metals and the rusting of iron**
- **know the conditions that promote rusting**
- **know methods used to delay or prevent rusting.**

How much do you already know?
Work out your score on page 116.

Test yourself

1 a) What conditions does iron need in order to rust? [3]
   b) What is the first step in the rusting of iron? [1]
   c) What reagent can be used to test for the rusting of iron in its early stages? [1]
   d) How is rusting detected? [1]

2 What is 'galvanising'? How does it protect iron from rusting? [4]

3 Explain what is meant by 'sacrificial protection.' How can it be used for underground pipes?

4

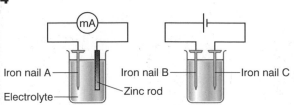

Iron nail A
Electrolyte
Zinc rod
Iron nail B
Iron nail C

a) Explain why **(i)** iron nail A does not rust and **(ii)** iron nail C does not rust. [6]
b) Write the ion–electron equation for the process taking place at iron nail B. [1]

12.1 Corrosion of metals

★ **Copper**: the green roofs you see on some buildings are of copper, which has corroded in the air to copper carbonate hydroxide, $Cu(OH)_2.CuCO_3$.

★ **Aluminium**: as soon as a fresh surface of aluminium meets the air, it is corroded to form a thin film of the oxide, which prevents air from reaching the metal below.

★ **Chromium** forms a protective oxide layer in the same way as aluminium. Stainless steel cutlery is made of a chromium–steel alloy (mixture).

★ **Nickel** forms a protective oxide layer as soon as a fresh surface of nickel meets the air. Nickel-plated steels are very useful.

★ **Lead** water pipes were used for centuries. However, water attacks lead slowly to form soluble lead compounds.

★ **Zinc** corrodes quickly in air to form a film of zinc carbonate. This protects the zinc beneath from further attack. Iron can be coated with zinc (**galvanised**) to protect it from rusting.

Although these metals corrode at different speeds, the reactions that take place are of the same kind. Metal atoms are oxidised to metal ions:

OXIDATION-
REDUCTION
Pages 70–71.

$$M \rightarrow M^{2+} + 2e^-$$

Rusting of iron and steel

The corrosion of iron and steel is called rusting. Rust has the formula $Fe_2O_3.nH_2O$, where n, the number of water molecules in the formula, varies.

The combination of reagents that attacks iron is water, air and acid. The carbon dioxide in the air provides the acidity. If the water contains salts, the speed of rusting is increased. In a warm climate, rusting is more rapid than at lower temperatures.

The first step in rusting is:

iron atoms → iron(II) ions + electrons

$$Fe(s) \rightarrow Fe^{2+}(aq) + 2e^-$$

Iron(II) ions can be further oxidised to iron(III) ions:

$$Fe^{2+}(aq) \rightarrow Fe^{3+}(aq) + e^-$$

The electrons which are released react with oxygen and water to form hydroxide ions, OH^-:

$$O_2(aq) + 2H_2O(l) + 4e^- \rightarrow 4OH^-(aq)$$

Iron(III) hydroxide, $Fe(OH)_3$, forms and is converted into rust, $Fe_2O_3.nH_2O$.

12.2 Experiments with a rust indicator

The wastage of iron through rusting is an expensive nuisance. Some methods of protecting iron from rusting involve giving it a protective coat of oil, grease, paint or plastic to keep air and water out (see table on page 86). Experiments to test these methods of protection normally take a long time. There is, however, a rust indicator which enables us to do rusting experiments more quickly. The green rust indicator called **ferroxyl** reacts with iron(II) ions to give a blue colour and with hydroxide ions to give a pink colour. In this way, the indicator can detect the early stages of rusting. The figure below shows petri dishes filled with a solution of ferroxyl indicator in agar gel. The gel is used because the colours that develop spread out less than they do in aqueous solution. Iron nails and iron wool are left on the gel for one week.

Such experiments show that:

- Iron wool rusts faster than iron nails.
- Nails rust faster at the tip and the head, which have been cut and hammered by a machine.
- Iron rusts faster in tap water than distilled water, and faster in a solution of salts (e.g. sea water) than in tap water.
- A layer of grease slows down rusting.
- A coat of paint will stop nails from rusting, as long as there are no chips in the paint.
- Galvanising (zinc-plating) is very effective at stopping rusting.
- Attaching a strip of magnesium, zinc, or another metal above iron in the reactivity series, prevents rusting. The more reactive metal corrodes instead. The protection stops when all the protective metal has corroded. Attaching a strip of tin or another metal below iron in the reactivity series speeds up rusting.

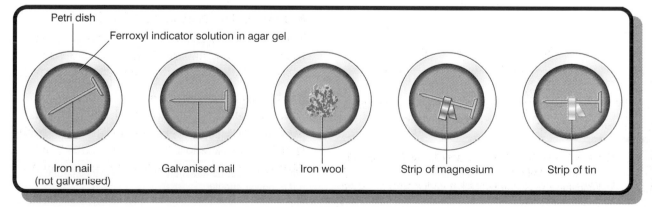

Petri dish

Ferroxyl indicator solution in agar gel

| Iron nail (not galvanised) | Galvanised nail | Iron wool | Strip of magnesium | Strip of tin |

Experiments with a rust indicator

12.3 Rusting as a redox reaction

Rusting is an oxidation–reduction reaction (see pages 70–71). The flow of electrons away from iron towards carbon is demonstrated in the diagram opposite. The ferroxyl indicator shows that Fe^{2+} ions are formed at the iron electrode and OH^- ions at the carbon electrode.

Do you understand it? What happens if the carbon rod is replaced by a) a strip of magnesium or b) a strip of tin? (Answer below.)

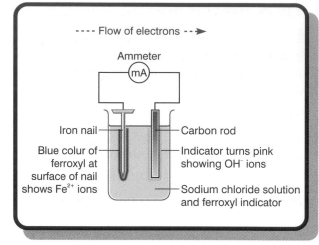

- - - - Flow of electrons - - ➤

Ammeter

mA

Iron nail — Carbon rod

Blue colur of ferroxyl at surface of nail shows Fe^{2+} ions — Indicator turns pink showing OH^- ions

Sodium chloride solution and ferroxyl indicator

Cell to demonstrate rusting as a redox reaction

Answer: **a)** Electrons flow from magnesium to iron. Iron does not rust. **b)** Electrons flow from tin to iron. Iron rusts faster than in the iron–carbon cell.

12.4 Rust prevention

method	where used	comment
1 a coat of paint	large objects, e.g. ships and bridges	If the paint is scratched, the iron beneath it starts to rust.
2 a film oil or grease	moving parts of machinery	The protective film must be renewed.
3 a coat of plastic	kitchen equipment	Lasts as long as the plastic remains intact.
4 a coat of metal **a)** chromium plating	trim on cars, cycle handlebars, taps	Applied by electroplating, decorative as well as protective.
b) galvanising (zinc plating)	galvanised steel girders are used in buildings	Even if the layer of zinc is scratched, the iron underneath does not rust. Zinc cannot be used for food cans because zinc and its compounds are poisonous.
c) tin plating	food cans	If the layer of tin is scratched, the iron beneath it rusts.
5 stainless steel	cutlery, car accessories	Steels containing chromium (10–25%) or nickel (10–20%) do not rust.
6 sacrificial protection	ships	Bars of zinc attached to the hull of a ship corrode and protect the ship from rusting.
	underground pipes	Bags of magnesium scrap attached to underground iron pipes corrode in preference to the pipes. The scrap must be replaced from time to time.
7 an applied negative potential	car bodies	The article must be attached to the negative electrode of a d.c. power supply.

Rust prevention

12.5 Sacrificial protection

Metals which are higher in the reactivity series than iron are more easily oxidised than iron. They are more ready than iron to provide electrons by the reaction:

$$M(s) \rightarrow M^{2+}(aq) + 2e^-$$

These metals corrode in preference to iron. They are used to protect iron structures such as underground pipes and ships. The two metals do not have to be in direct contact. If they are connected by a wire the results are the same. Bags of magnesium scrap connected to underground pipes protect against rusting. Magnesium corrodes while the iron remains intact. This is called **sacrificial protection**. When the magnesium has corroded it must be replaced. Zinc bars attached to the hulls of ships offer sacrificial protection in the same way.

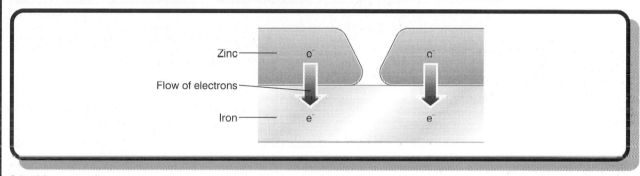

Galvanising protects iron

The coating of iron with zinc is called **galvanising**. Zinc provides a barrier to air and water and prevents corrosion. If the layer of zinc becomes scratched, protection continues because zinc corrodes in preference to iron.

The 'tin cans' used to preserve food are made of iron coated with tin because tin does not corrode. The tin coating may be scratched through mishandling to expose the iron beneath. The can then rusts faster than iron alone. Since tin is below iron in the electrochemical series the presence of tin enables a cell to be set up:

$$Fe(s) \rightarrow Fe^{2+}(aq) + 2e$$

$$Sn^{2+}(aq) + 2e^- \rightarrow Sn(s)$$

By accepting electrons from iron atoms, tin ions hasten the rusting of iron.

The loss of electrons is oxidation. One method of preventing the loss of electrons is to attach iron to the negative terminal of a battery. Then the tendency of iron atoms to lose electrons is counteracted by the supply of electrons from the battery. This method is adopted in vehicles, the body work being connected to the negative terminal of the battery.

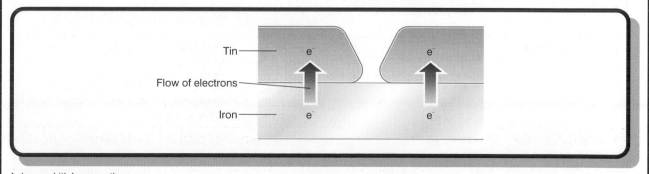

A damaged 'tin' can rusting

12

round-up

How much have you improved?
Work out your improvement index on page 116–117.

1 The prevention of rusting in car bodies is important. Give four methods that are employed. Explain how each method works. [10]

2 Suggest an experiment to compare the abilities of tin and zinc to affect the speed at which iron rusts. [4]

3 Many foods are preserved in 'tin' cans. If the tin coating is scratched the iron underneath rusts rapidly. Why does this happen? [2]

4 Iron nails are electroplated with nickel to make them rust-resistant.
a) Why is electroplating a good method of applying a nickel coating? [2]
b) Why are the nails to be plated connected to the negative terminal of the power supply? [1]
c) Write the equation for the process at the negative electrode. [3]
d) If the positive electrode is a piece of nickel write the equation for the process at the positive electrode. [3]
e) If nickel sulphate is used as electrolyte, does the concentration of nickel sulphate increase or decrease or stay the same? Explain your answer. [2]

5 One iron nail is placed in each of a number of beakers filled with water. Which of the following will delay rusting?
A attaching a piece of magnesium ribbon to the iron nail
B attaching a piece of tin foil to the iron nail
C connecting the nail to the positive terminal of a power supply
D connecting the nail to the negative terminal of a power supply
E adding sodium nitrate to the water
F adding sodium hydroxide to the water. [3]

6 Which will rust more rapidly, nail A or nail B? Explain your answer. [5]

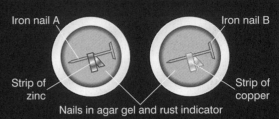

Iron nail A   Iron nail B
Strip of zinc
Strip of copper
Nails in agar gel and rust indicator

7 Say which method you would choose to delay rusting of each of the following objects.
a) an iron rail along the seafront
b) an oil tanker above the water line
c) an oil tanker below the water line
d) a fire escape outside an apartment block
e) the base plate of an electric iron
f) a knife and fork
g) an underground gas pipe
h) bicycle handlebars
i) bicycle brake cables
j) a bicycle chain
k) a food can
l) a car body
m) a car hub cap
n) a bridge. [14]

8 Fraser is deciding whether to buy **a)** a steel car exhaust for £94 which will last for about 2 years before it has to be replaced because of rusting or **b)** a stainless steel exhaust system for £180 which will last for about 6 years before it needs replacing. Help Fraser to decide. Calculate how much it will cost him per year to buy **a)** and **b)**. [4]

9 Some iron nails are placed in dishes containing agar gel and rust indicator. The figure shows the results after one week. What two conclusions can you draw?

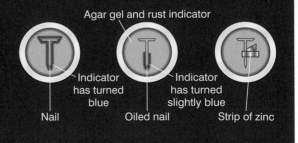

Agar gel and rust indicator
Indicator has turned blue
Indicator has turned slightly blue
Nail   Oiled nail   Strip of zinc

Plastics and synthetic fibres

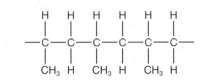

preview

At the end of this topic you will know:

- **how plastics and synthetic fibres are derived from oil**
- **the difference between thermosoftening and thermosetting plastics**
- **some of the many uses of plastics**
- **how waste plastics are disposed of.**

ALKENES Page 41.
MIND MAP
Page 126.

How much do you already know?
Work out your score on page 117.

Test yourself

1 Which of the following is the material from which most plastics are derived:
coal, oil, peat, natural gas, wood? [1]

2 Cotton is a natural fibre. Nylon is a synthetic fibre.
  a) What does the word 'synthetic' mean? [1]
  b) Why is it necessary to use synthetic fibres? [1]
  c) What are the advantages of **(i)** cotton over nylon **(ii)** nylon over cotton? [2]

3 Give the meanings of the terms monomer, polymer and polymerise. Illustrate your answer by referring to ethene. [4]

4 a) How do a thermosoftening plastic and a thermosetting plastic differ in behaviour? [2]
  b) How does the difference in structure explain the difference in behaviour? [2]
  c) Give two examples of uses for which a thermosetting plastic is chosen and two for which a thermosoftening plastic is chosen. [4]

5 a)

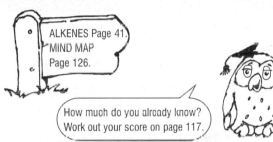

The above formula is part of a polymer molecule. Write the formula of the monomer and give its name. [2]
  b) $CH_2=CHCl$ is a monomer. Draw the structural formula of the polymer which it forms. [1]
  c) Give **(i)** the chemical name **(ii)** the trade name **(iii)** two uses for this polymer. [4]

6 What is meant by the term 'non-biodegradable'? Why is the non-biodegradability of plastics a problem? [2]

13.1 Properties of plastics

Plastics are mainly **synthetic** (man-made) materials. Rubber is a natural plastic. Plastics are being used in place of natural materials for many purposes. You will be able to think of many examples of plastics being used instead of wood, china, and glass.

Plastics are
- strong
- low in density
- good insulators of heat and electricity
- resistant to attack by chemicals
- smooth
- able to be moulded into different shapes.

There are two kinds of plastics, **thermoplastics** and **thermosetting plastics**. The difference is shown in the diagram below.

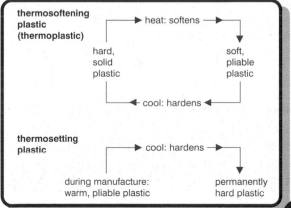

The reason for the difference in behaviour is a difference in structure, as shown.

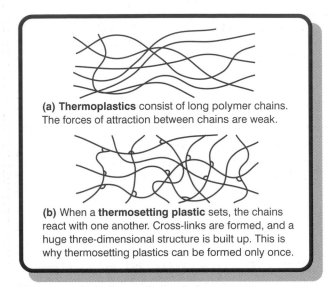

(a) **Thermoplastics** consist of long polymer chains. The forces of attraction between chains are weak.

(b) When a **thermosetting plastic** sets, the chains react with one another. Cross-links are formed, and a huge three-dimensional structure is built up. This is why thermosetting plastics can be formed only once.

The structure of (a) a thermosoftening plastic
(b) a thermosetting plastic

Moulding of thermoplastics can be a **continuous process**: solid granules of the plastic are fed into one end of the moulding machine, softened by heat and then moulded to come out of the other end in the shape of tubes, sheets or rods. It is easy to manufacture coloured articles by adding a pigment to the plastic.

The moulding of thermosetting plastics is a **batch process**. The monomer is poured into the mould and heated. As it polymerises, the plastic solidifies and a press forms it into the required shape while it is setting.

Both types of plastic have their advantages. A material used for electrical fittings and counter tops must be able to withstand high temperatures without softening. For these purposes, 'thermosets' are used.

Sometimes gases are mixed with softened plastics to make low density plastic foam for use in car seats, thermal insulation, sound insulation and packaging. Plastics can be strengthened by the addition of other materials; for example the composite material **glass fibre-reinforced plastic** is used for the manufacture of boat hulls and car bodies.

13.2 Source of plastics

Crude oil is fractionally distilled to give a number of fractions of different boiling point ranges (see page 33). There is more demand for the fractions called gasoline and kerosene than for the higher boiling point fractions. These fractions are **cracked** to give more gasoline and kerosene and also alkenes (see page 41). Alkenes are hydrocarbons with a double bond between two carbon atoms, e.g. ethene, $H_2C=CH_2$.

13.3 Poly(alkenes)

Alkenes take part in addition reactions (see page 41). One of these is **addition polymerisation**. In this reaction many molecules of the **monomer**, e.g. ethene, join together (**polymerise**) to form the **polymer**, e.g. poly(ethene).

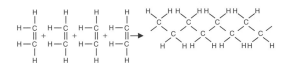

The conditions needed are:

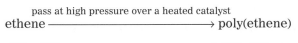

$$\text{ethene} \xrightarrow{\text{pass at high pressure over a heated catalyst}} \text{poly(ethene)}$$

$$n\text{CH}_2=\text{CH}_2 \longrightarrow (-\text{CH}_2-\text{CH}_2-)_n$$

In poly(ethene), n is between 30 000 and 40 000. Poly(ethene) is used for making plastic bags, for kitchenware (buckets, bowls, etc.), for laboratory tubing and for toys. It is flexible and difficult to break. Polymers of alkenes are called poly(alkenes).

some poly(alkenes) and their uses	
alkene	**use**
Poly(ethene); trade name Polythene monomer $C=C$ (with H, H on each) polymer $(-C-C-)_n$	Polythene is used to make plastic bags. High density polythene is used to make kitchenware, laboratory tubing and toys.
Poly(chloroethene); trade name PVC monomer $C=C$ (with Cl, H / H, H) polymer $(-C-C-)_n$	PVC is used to make plastic bottles, wellingtons and raincoats, floor tiles, insulation for electrical wiring, gutters and drainpipes.
Poly(propene) monomer $C=C$ (with H, CH_3 / H, H) polymer $(-C-C-)_n$	Poly(propene) is resistant to attack by chemicals and does not soften in boiling water. It can be used to make hospital equipment which must be sterilised. Poly(propene) is drawn into fibres and used to make ropes and fishing nets.
Poly(tetrafluoroethene); trade names PTFE and Teflon monomer $C=C$ (with F, F / F, F) polymer $(-C-C-)_n$	PTFE is a hard, waxy plastic which is not attacked by most chemicals. Few substances can stick to its surface. It is used to coat non-stick pans etc.
Perspex monomer $C=C$ (with H, CO_2CH_3 / H, CH_3)	Perspex is an important plastic because it is transparent and can be used instead of glass. It is more easily moulded than glass and less easily shattered.
Polystyrene monomer $C=C$ (with H, H / H, C_6H_5)	Polystyrene is a hard, brittle plastic used for making construction kits. Polystyrene foam is made by blowing air into the softened plastic. It is used for making insulating containers and packaging of fragile goods.

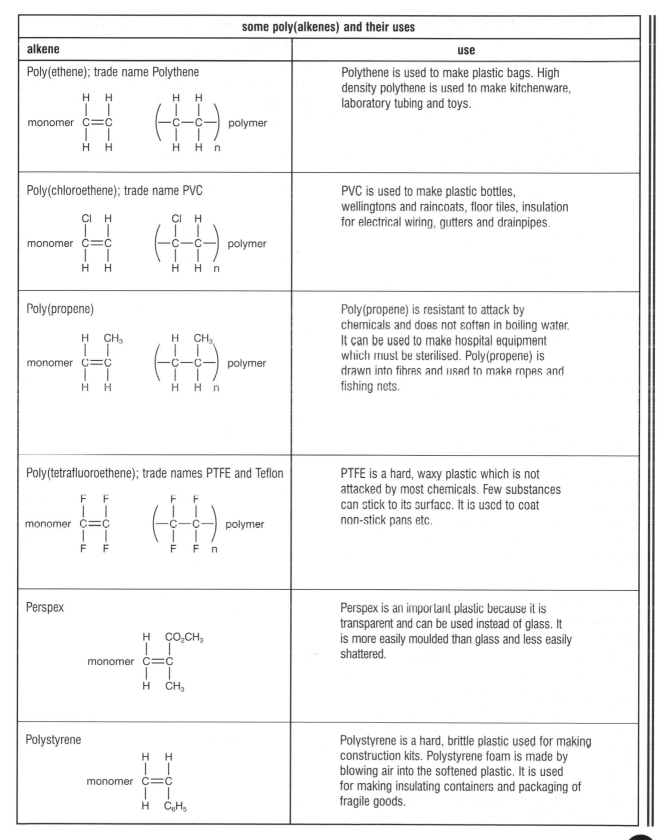

13.4 Condensation polymers

The chief **thermosetting plastics (thermosets)** are not poly(alkenes). They are **condensation polymers**. In **condensation polymerisation** many molecules of monomer add to form one molecule of polymer and many molecules of water. Some useful thermosets are described in the table below.

some thermosetting condensation polymers and their uses	
polymer	use
Bakelite	Resistant to chemical attack, high m.p., insoluble, brown. Used in electrical switches, plugs and sockets.
Urea-methanal	Similar to bakelite but colourless so it can be coloured with pigments. Used for electrical fittings.
Melamine	Used for kitchen surfaces.
Epoxy resins	Used as glues.
Polyester resins	Used in glass-reinforced plastics.
Polyurethanes	Used in varnishes.

13.5 Synthetic fibres

These fibres are made by forcing molten plastic through a fine hole. As the fibre cools and solidifies, it is stretched to align its molecules along the length of the fibre. This is why fibres have great tensile strength along the length of the fibre.

name	uses
Nylon	High tensile(stretching) strength, high m.p. (200 °C). Used in textile fibres for making clothes and for ropes, fishing lines and nets.
Terylene®	Used in textile fibres. A disadvantage of Terylene® and nylon is that they do not absorb water well and, therefore, clothes made of these fibres do not allow perspiration to evaporate through them. They are often mixed with natural fibres, such cotton and wool.

Two synthetic fibres and their uses

13.6 Some drawbacks to plastics

Streets, parks and beaches are littered with discarded plastic bags and food containers. Landfill sites are being filled with tonnes of plastic waste which never rots. Most plastics are non-biodegradable. They are synthetic materials. Natural materials such as wood and paper are decomposed by micro-organisms in the soil. There are no micro-organisms in the soil to decompose plastics.

Many buildings are insulated with plastic foam or furnished with plastic materials. Plastics have low ignition temperatures, and fires can spread rapidly when plastics burn. Some plastics form toxic combustion products, such as hydrogen chloride (from PVC) and hydrogen cyanide (from polyurethanes). Builders are not allowed to use such plastics.

Biodegradable plastics

Chemists are working on the problem of disposal of plastics. They have invented some biodegradable plastics.

★ The ICI biopolymer called PHB is made by certain bacteria. Micro-organisms in the soil, in river water and in the body can break it down within 9 months.

★ A photodegradable polymer can be incorporated in plastic food containers. Exposed to sunlight for 60 days, the containers break down into dust particles.

★ A material made of poly(ethene) and starch can be used for carrier bags. When the material is buried, micro-organisms convert the starch into carbon dioxide and water, and in time the material disintegrates. The cost at present is about twice that of a regular plastic bag.

Recycling of plastics

When plastics are recycled, there is a saving on raw materials and, at the same time the problem of disposing of non-biodegradable plastics is reduced. At present, the recycling of plastics is still at the experimental stage Problems still to be solved include the difficulty of separating different types of plastic, and the difficulty of removing additives.

Oil: a fuel and a source of petrochemicals

The number of plastics and the uses found for them is constantly growing. The raw materials used in their manufacture come from oil. The Earth's resources of oil will not last for ever. It seems wasteful to burn oil as fuel when we need it to make plastics and other petrochemicals.

round-up

How much have you improved?
Work out your improvement index on page 117–118.

1 State the advantage that plastic has
   a) over china for making cups and saucers
      and dolls   [1]
   b) over lead for making toy farmyard animals
      and soldiers   [1]
   c) over glass for making motorbike windscreens.   [1]
   d) the formula for the monomer of the plastic used for
      windscreens is given in the table on page 91. Write the
      formula for the polymer.   [1]

2 a) What does the word 'plastic' mean?   [2]
   b) There are two big classes of plastics, which behave
      differently when heated. Name the two classes.
      Describe the difference in behaviour. Say how this
      difference is related to the molecular nature of
      the plastics.   [5]

3 a) Write the structural formulae for (i) tetrafluoroethene
      (ii) poly(tetrafluoroethene), PTFE.   [2]
   b) Give two uses for PTFE.   [2]

4 What properties are needed for the materials used for
   (i) electrical fittings and (ii) kitchen work surfaces?
   Suggest one material for each of these uses.   [4]

5 Millions of plastic bags are discarded after one or two
   hours' use. Many plastic bags are made of
   poly(ethene).
   a) Explain how poly(ethene) is obtained from
      petroleum.   [3]
   b) How long did petroleum take to form?   [1]
   c) Can it be replaced?   [1]

6 a) What is meant by the statement that plastics are
      non-biodegradable?   [1]
   b) Why can this be a disadvantage?   [1]
   c) Mention an advantage of burning waste plastics.
                                                         [1]
   d) What toxic substance is produced by all plastics if
      they burn in insufficient air?   [1]
   e) What toxic substances are produced by burning
      PVC?   [2]
   f) Suggest an alternative to burning waste plastics.
                                                         [1]

Well done if you've improved. Don't worry
if you haven't. Take a break and try again.

Fertilisers

preview

At the end of this section you will:

- **understand why NPK fertilisers are important**
- **know about the manufacture and reactions of ammonia**
- **know about the manufacture of nitric acid and NPK fertilisers.**

MIND MAP
Page 127.

How much do you already know?
Work out your score on page 118.

14.1 Plant nutrients

Plants need

- carbon, hydrogen and oxygen compounds, which are in plentiful supply
- small quantities of **trace elements** (iron, manganese, boron, copper, cobalt, molybdenum, zinc), which are present in most soils
- fairly large quantities of compounds containing nitrogen, phosphorus, potassium, calcium, magnesium, sulphur and, for some crops, sodium.

NITROGEN CYCLE
Page 96.

Test yourself

1 Unlike most plants, clover can convert nitrogen in the air into nitrogen compounds. How does it do this? [3]

2 What do nitrifying bacteria do? [3]

3 What do denitrifying bacteria do? [3]

4 NPK fertilisers are widely used. What types of compound do they contain? [3]

5 Some of the nitrate content of fertilisers washes into lake water. What problems does this cause? [3]

6 What is the percentage by mass of nitrogen in urea, CON_2H_4? (Relative atomic masses: H = 1, C = 12, N = 14, O = 16) [2]

7 In the Haber process for manufacturing ammonia, what conditions are used? [3]

8 Ammonium nitrate is used as a fertiliser.
   a) Show in a flow diagram how it is made, starting from nitrogen and hydrogen. [3]
   b) What is **(i)** the advantage **(ii)** the disadvantage of the high solubility of ammonium nitrate in its use as a fertiliser? [2]

9 Name three raw materials needed for the manufacture of NPK fertilisers. [3]

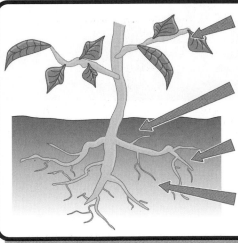

Potassium: potassium compounds assist photosynthesis. NPK fertilisers contain potassium chloride.

Nitrogen: nitrogenous fertilisers increase both the size of the crop and the protein content of the plants. Plants can absorb nitrates from the soil. Ammonium salts can also be used as fertilisers, because ammonium salts are converted into nitrates by microorganisms in the soil.

Phosphorus: phosphates stimulate root development. The soluble fertiliser ammonium phosphate supplies phosphorus to crops.

Sodium: a few crops, e.g. sugar beet, grow better if common salt is added as a fertiliser.

Some of the elements that plants need

Fertilisers that supply nitrogen, phosphorus and potassium are called **NPK fertilisers**. They are used in huge quantities: about 20 million tonnes a year worldwide. Fertilisers cost about £80 per tonne and are a large item in a farmer's budget. The use of excessive fertilisers is wasteful and often causes pollution.

POLLUTION
Page 36–38.

14.2 Ammonia

The first step in the manufacture of nitrogenous fertilisers is the manufacture of ammonia by the Haber process. A test for ammonia, which is also a test for hydrogen chloride, is shown in the diagram opposite, and the reactions of ammonia are summarised in the concept map below.

HABER PROCESS
Pages 97.

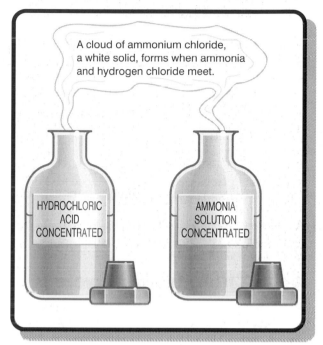

A cloud of ammonium chloride, a white solid, forms when ammonia and hydrogen chloride meet.

HYDROCHLORIC ACID CONCENTRATED

AMMONIA SOLUTION CONCENTRATED

The reaction between ammonia and hydrogen chloride

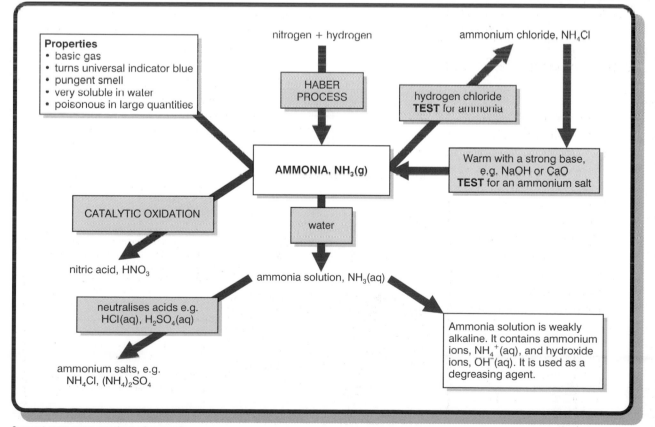

Properties
- basic gas
- turns universal indicator blue
- pungent smell
- very soluble in water
- poisonous in large quantities

nitrogen + hydrogen

ammonium chloride, NH_4Cl

HABER PROCESS

hydrogen chloride **TEST** for ammonia

AMMONIA, $NH_3(g)$

Warm with a strong base, e.g. NaOH or CaO **TEST** for an ammonium salt

CATALYTIC OXIDATION

water

nitric acid, HNO_3

ammonia solution, $NH_3(aq)$

neutralises acids e.g. $HCl(aq)$, $H_2SO_4(aq)$

Ammonia solution is weakly alkaline. It contains ammonium ions, $NH_4^+(aq)$, and hydroxide ions, $OH^-(aq)$. It is used as a degreasing agent.

ammonium salts, e.g. NH_4Cl, $(NH_4)_2SO_4$

Concept map: some reactions of ammonia

14

14.3 Nitrogen

Nitrogen is a gas which does not take part in many chemical reactions. It combines with hydrogen to form ammonia. This reaction is the basis of the fertiliser industry; see opposite. Many uses of nitrogen arise from its lack of reactivity.

★ Liquid nitrogen is used in the fast-freezing of foods.

★ Many foods are packed in an atmosphere of nitrogen to prevent oils and fats in the foods from being oxidised to rancid products.

★ Oil tankers, road tankers and grain silos are flushed out with nitrogen as a precaution against fire.

14.4 The nitrogen cycle

Nitrogen circulates from air to soil to living things and back again in a process called the **nitrogen cycle**, shown below.

14.5 Ammonia and fertilisers

Ammonia

The nitrogen in the air is used to make nitrogenous fertilisers. Chemical methods of fixing nitrogen are more costly than bacterial methods. Under the conditions of the **Haber process** (named after the chemist Fritz Haber) nitrogen will combine with hydrogen.

$$\text{nitrogen} + \text{hydrogen} \rightleftharpoons \text{ammonia}$$

$$N_2(g) + 3H_2(g) \rightleftharpoons 2NH_3(g)$$

The reaction is reversible: some of the ammonia formed dissociates into nitrogen and hydrogen. The product is a mixture of nitrogen, hydrogen and ammonia. Two factors increase the percentage of ammonia in the mixture: a **high pressure** and a **low temperature**. However, the reaction is very slow at a low temperature, and industrial plants use a compromise temperature and a catalyst to speed up

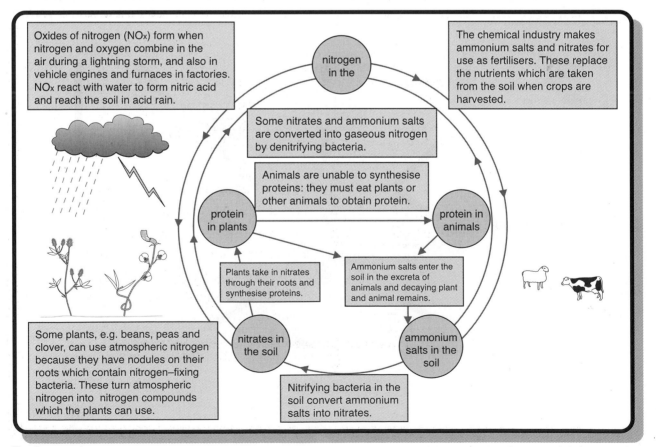

Oxides of nitrogen (NOx) form when nitrogen and oxygen combine in the air during a lightning storm, and also in vehicle engines and furnaces in factories. NOx react with water to form nitric acid and reach the soil in acid rain.

The chemical industry makes ammonium salts and nitrates for use as fertilisers. These replace the nutrients which are taken from the soil when crops are harvested.

nitrogen in the

Some nitrates and ammonium salts are converted into gaseous nitrogen by denitrifying bacteria.

Animals are unable to synthesise proteins: they must eat plants or other animals to obtain protein.

protein in plants

protein in animals

Plants take in nitrates through their roots and synthesise proteins.

Ammonium salts enter the soil in the excreta of animals and decaying plant and animal remains.

nitrates in the soil

ammonium salts in the soil

Some plants, e.g. beans, peas and clover, can use atmospheric nitrogen because they have nodules on their roots which contain nitrogen–fixing bacteria. These turn atmospheric nitrogen into nitrogen compounds which the plants can use.

Nitrifying bacteria in the soil convert ammonium salts into nitrates.

The nitrogen cycle

the reaction, as shown in the flow diagram. The ammonia made by the Haber process can be oxidised to nitric acid.

14.6 NPK fertilisers

Ammonia solution can be used as a fertiliser. However, it is more common to use the solid fertilisers ammonium nitrate, ammonium sulphate and ammonium phosphate.

Mixtures of ammonium nitrate, ammonium phosphate and potassium chloride contain the elements nitrogen, phosphorus and potassium, which are essential for plant growth and are sold as **NPK fertilisers**.

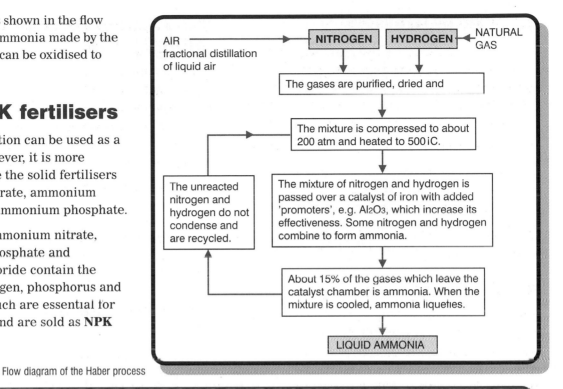

Flow diagram of the Haber process

Flow diagram for the manufacture of NPK fertilisers

Pollution by fertilisers

Fertilisers: when an excess of fertiliser is used, some of it is not absorbed by the crop. Rain washes it out of the soil, and it accumulates in groundwater. The water industry uses groundwater as a source of drinking water. There is concern that nitrates in drinking water can lead to the formation of nitrosoamines, compounds which cause cancer.

Fertiliser which plants fail to absorb may be carried into the water of a loch, where it nourishes the growth of algae and water plants. This accidental enrichment of the water causes algae to form a thick mat of **algal bloom**, and weeds flourish. When algae die and decay, they use up dissolved oxygen. The fish in the loch are deprived of oxygen and die. The loch becomes a 'dead' loch. This process is called **eutrophication**.

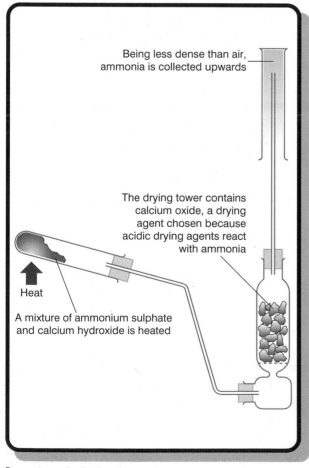

Preparation and collection of ammonia

14.7 Laboratory preparation of ammonia

Ammonia is a weak base. A strong base, such as sodium hydroxide, calcium oxide or calcium hydroxide, will drive the weak base ammonia out of its salts. Gentle heating speeds up the reaction. In the example shown below,

ammonium + calcium → ammonia + calcium + water
 sulphate   hydroxide   sulphate

$(NH_4)_2SO_4(s) + Ca(OH)_2(s) \rightarrow 2NH_3(g) + CaSO_4(s) + 2H_2O(l)$

This reaction can be used as a **test for ammonium salts**. All ammonium salts give ammonia when heated with a strong base.

14.8 Manufacture of nitric acid

Nitric acid is made by the reaction of nitrogen dioxide and oxygen with water.

$4NO_2(g) + O_2(g) + 2H_2O(l) \rightarrow 4HNO_3(l)$

The formation of nitrogen dioxide by the combination of nitrogen and oxygen is **endothermic**; energy is required.

$N_2(g) + 2O_2(g) \rightarrow 2NO_2(g)$

The reaction takes place in the atmosphere during lightning storms and in vehicle engines. Nitrogen dioxide is not made by this method industrially because of the high cost of raising the reactants to the required temperature. Instead, it is made by the catalytic oxidation of ammonia in the **Ostwald process**. In the presence of a platinum catalyst, ammonia is oxidised by air to nitrogen monoxide, NO. This reaction is **exothermic**. Heat must be supplied to get the reaction started, but once it starts the reaction liberates enough heat to keep it going without further heating. Nitrogen monoxide is immediately oxidised to nitrogen dioxide, NO_2.

$$\text{Air} + \text{ammonia} + \text{platinum} \xrightarrow{} \overset{\text{air}}{\text{nitrogen}} \rightarrow \overset{\text{air}}{\text{nitrogen}} \rightarrow \text{nitric acid } HNO_3$$

$$\begin{array}{ccc} \text{catalyst} & \text{monoxide} & \text{dioxide }\text{\small water} \\ +\,800\,°C & NO & NO_2 \end{array}$$

The industrial oxidation of ammonia can be duplicated in the laboratory, e.g. in the apparatus shown in the diagram below.

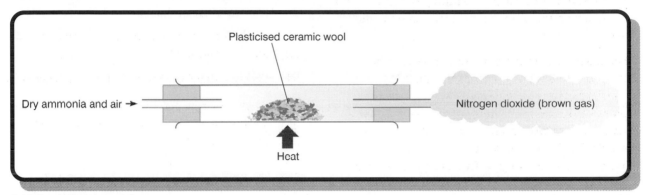

Catalytic oxidation of ammonia

14.9 Preparation of ammonium sulphate

Ammonium sulphate is used as a fertiliser. It is made industrially by the reaction between ammonia and sulphuric acid.

ammonia + sulphuric acid → ammonium sulphate
$$2NH_3(aq) + H_2SO_4(aq) \rightarrow (NH_4)_2SO_4(aq)$$

To make ammonium sulphate in the laboratory:

1 Add ammonia solution slowly to dilute sulphuric acid, with stirring.

2 From time to time, test a drop of the solution with universal indicator paper.

3 When the universal indicator paper turns green, all the acid has been neutralised, and there is a slight excess of ammonia present.

4 Heat the solution to drive off the excess of ammonia and evaporate some of the water. When the solution is sufficiently concentrated, crystals of ammonium sulphate form.

14.10 The reaction between a base and an ammonium salt

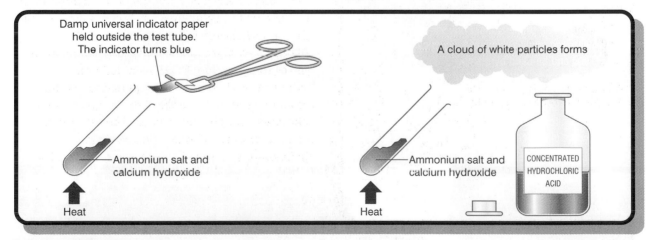

Testing for ammonium salt

round-up

How much have you improved?
Work out your improvement index on page 118.

1 Find the percentage by mass of nitrogen in each of the following fertilisers.
  a) ammonia, NH_3 **b)** ammonium nitrate, NH_4NO_3
  c) ammonium sulphate, $(NH_4)_2SO_4$ **d)** urea, CON_2H_4
  [4]

2 Write the word equation and chemical equation for the reaction between ammonia and hydrogen chloride on page 109. [3]

3 Write a chemical equation for this example of the displacement of ammonia from one of its salts: [3]

$$\frac{\text{ammonium}}{\text{chloride}} + \frac{\text{sodium}}{\text{hydroxide}} \rightarrow \text{ammonia} + \frac{\text{sodium}}{\text{chloride}} + \text{water}$$

4 Suggest what **A**, **B**, **C**, **D** and **E** might be.

  A is a neutral, rather unreactive gas which reacts with hydrogen under pressure.

  The indicator **B** is turned blue by ammonia.

  The gas **C** reacts with ammonia to form a white solid.

  Ammonium sulphate gives ammonia when it is warmed with a solution of **D**.

  Gas **E** forms an explosive mixture with air but combines with **A** with difficulty. [5]

5 a) What nutritious compounds do plants make from nitrates? [1]
  b) How can atmospheric nitrogen be used as a fertiliser by some plants? [2]
  c) Name one process that converts atmospheric nitrogen into a compound which can enter the soil and be used by plants. [1]
  d) Explain why natural sources of nitrogen will not support the repeated growing of crops on the same land. [2]
  e) Why are ammonium salts used as fertilisers when plants cannot absorb them? [3]

NITROGEN CYCLE
Page 96.

6 A ammonia **B** oxygen **C** carbon dioxide **D** nitrogen **E** nitrogen monoxide.
  Choose from these gases the one that:
  a) relights a glowing splint **b)** turns limewater milky
  c) gives an alkaline solution **d)** is formed in lightning storms **e)** forms the highest percentage by volume of air. [5]

7 What is meant by 'fixing nitrogen'? State one natural method by which nitrogen is fixed and one industrial method. [3]

8 The manufacture of fertilisers depends on ammonia:
  Ammonia + oxygen → nitrogen dioxide → nitric acid → synthetic fertilisers
  a) What are the starting materials for the manufacture of ammonia? [2]
  b) What conditions are employed in the manufacture of ammonia? [3]
  c) Why is nitric acid made from ammonia and not from nitrogen? [2]

9 Balanced fertilisers contain nitrogen, phosphorus and potassium. The table lists four substances which are used in fertilisers.

name	formula
basic slag	$(CaO)_5.P_2O_5.SiO_2$
ammonium sulphate	$(NH_4)_2SO_4$
urea	$CO(NH_2)_2$
sylvite	KCl

  a) Which of the substances in the table contains phosphorus? [1]
  b) Which three substances would you need to use to obtain a balanced fertiliser? [3]

Well done if you've improved. Don't worry if you haven't. Take a break and try again.

Carbohydrates

preview

At the end of this topic you will know about:

- **the carbon cycle**
- **the stucture and properties of carbohydrates**
- **the structure of alkanols**
- **ethanol (alcohol).**

How much do you already know?
Work out your score on page 118–119.

Test yourself

1 Name the substances **(i)** used by plants in photosynthesis **(ii)** formed when sugars are burned in the lab **(iii)** formed when sugars are respired in animals **(iv)** produced when glucose is fermented.

2 A starch **B** sucrose **C** glucose **D** cellulose **E** ethanol. Which substance **(i)** gives a positive result with Benedict's reagent **(ii)** gives a blue colour with iodine solution **(iii)** is obtained by fermentation of sugars **(iv)** has the formula $C_{12}H_{22}O_{11}$ **(v)** is the polysaccharide in plant cell walls? [5]

3 A carbon dioxide **B** methane **C** nitrogen **D** oxygen **E** water vapour.
Which of the gases listed is or are:
a) produced in photosynthesis
b) produced in the fermentation of glucose
c) inhaled by animals but not respired
d) burnt as fuel
e) formed when carbohydrates burn? [6]

4 Say how you could distinguish between:
a) sucrose and starch [3]
b) sucrose and glucose [3]
c) sucrose and maltose. [5]

5 The carbon cycle should operate so that the percentage of carbon dioxide in the air remains constant.
a) Mention two factors that can lead to an increase in the percentage of carbon dioxide in the air. [2]
b) Why is there concern over rising carbon dioxide levels? [2]

6 Glucose is formed in plant cells. Name:
a) the process involved
b) the gas which plants use in the process
c) the gas which is given out
d) the catalyst for the process
e) the substance formed in plant cells by polymerisation of glucose. [2]

7 A combustion of methane
B distillation of crude oil
C combination of nitrogen and hydrogen
D fermentation of glucose
E polymerisation of ethene
From the list of industrial processes choose that which is used in the manufacture of **(i)** plastics **(ii)** alcoholic drinks **(iii)** fertilisers **(iv)** lubricating oil. [4]

15.1 Carbon dioxide

The carbon cycle

Plants are able to make sugars by the process of **photosynthesis:**

$$\text{carbon dioxide} + \text{water} + \text{sunlight} \xrightarrow{\text{catalysed by chlorophyll}} \text{glucose} + \text{oxygen}$$

The energy of sunlight is converted into the energy of the chemical bonds in the sugar glucose.

Animals obtain energy by **cellular respiration:**

$$\text{glucose} + \text{oxygen} \rightarrow \text{carbon dioxide} + \text{water} + \text{energy}$$

The balance between the processes which take carbon dioxide from the air and those which put carbon dioxide into the air is called the **carbon cycle**. The cycle is shown below.

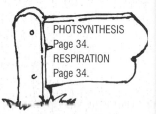

PHOTSYNTHESIS
Page 34.
RESPIRATION
Page 34.

15

Uses of carbon dioxide

★ Soft drinks are made by dissolving carbon dioxide in water under pressure and adding sugar and flavourings.

★ Solid carbon dioxide sublimes (turns into a vapour on warming). It is used as the refrigerant 'dry ice'.

★ Carbon dioxide is used in fire extinguishers because it does not support combustion and is denser than air.

Test for carbon dioxide

Carbon dioxide reacts with a solution of calcium hydroxide (limewater) to form a white precipitate of calcium carbonate.

$$\text{carbon dioxide} + \text{calcium hydroxide} \rightarrow \text{calcium carbonate} + \text{water}$$

$$CO_2(g) + Ca(OH)_2(aq) \rightarrow CaCO_3(s) + H_2O(l)$$

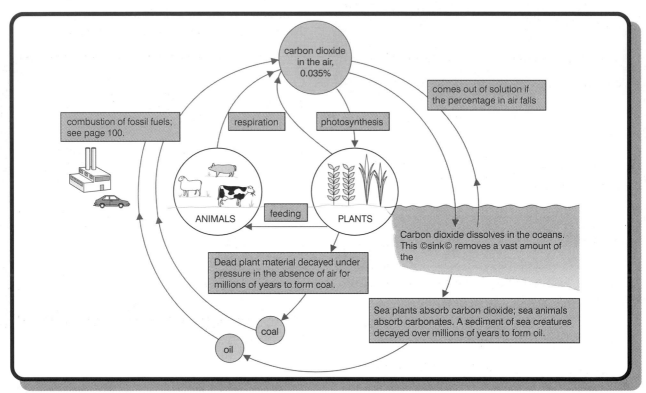

carbon dioxide in the air, 0.035%

comes out of solution if the percentage in air falls

combustion of fossil fuels; see page 100.

respiration

photosynthesis

ANIMALS

feeding

PLANTS

Carbon dioxide dissolves in the oceans. This ©sink© removes a vast amount of the

Dead plant material decayed under pressure in the absence of air for millions of years to form coal.

Sea plants absorb carbon dioxide; sea animals absorb carbonates. A sediment of sea creatures decayed over millions of years to form oil.

coal

oil

The carbon cycle

15.2 Carbohydrates

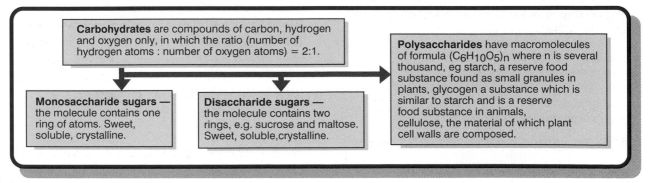

Carbohydrates are compounds of carbon, hydrogen and oxygen only, in which the ratio (number of hydrogen atoms : number of oxygen atoms) = 2:1.

Polysaccharides have macromolecules of formula $(C_6H_{10}O_5)_n$ where n is several thousand, eg starch, a reserve food substance found as small granules in plants, glycogen a substance which is similar to starch and is a reserve food substance in animals, cellulose, the material of which plant cell walls are composed.

Monosaccharide sugars — the molecule contains one ring of atoms. Sweet, soluble, crystalline.

Disaccharide sugars — the molecule contains two rings, e.g. sucrose and maltose. Sweet, soluble, crystalline.

Carbohydrates

Sucrose is formed from glucose and fructose:

glucose + fructose → sucrose + water

$$C_6H_{12}O_6(aq) + C_6H_{12}O_6(aq) → C_{12}H_{22}O_{11}(aq) + H_2O(l)$$

Maltose is formed from glucose:

glucose → maltose + water

$$2C_6H_{12}O_6(aq) → C_{12}H_{22}O_{11}(aq) + H_2O(l)$$

Model of a glucose molecule

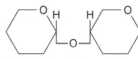

Shorthand formula for maltose

These reactions are examples of **condensation polymerisation**. Further condensation polymerisation of glucose results in the formation of starch

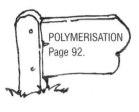

POLYMERISATION Page 92.

Carbohydrates have a vital function in all living organisms: they provide energy. In respiration, carbohydrates are oxidised to carbon dioxide and water with the release of energy. Living organisms also respire fats, oils and proteins.

The oxidation of glucose in **cellular respiration** is summarised by the equation:

$$C_6H_{12}O_6(aq) + 6O_2(g) → 6CO_2(g) + 6H_2O(l); energy$$
is released

Glucose does not make a good storage food because it is very soluble. For storage, plants convert glucose into starch and animals convert glucose into glycogen (see below). The reaction which takes place in plants is:

glucose → starch + water

$$nC_6H_{12}O_6(aq) → (C_6H_{10}O_5)_n(s) + nH_2O(l)$$

When cells need energy they convert starch and glycogen back into glucose.

Starch, glycogen and cellulose

Starch and **glycogen** are food substances. Starch is stored in plant cells and glycogen is stored in animal cells. The structure of starch is shown at the top of the next column. **Cellulose** is the substance of which plant cell walls are composed. All these substances are carbohydrates. They are **polysaccharides**, that is, their molecules contain a large number of glucose rings (several hundred).

Starch and glycogen make good food stores because they are only slightly soluble in water. They can remain in the cells of an organism without dissolving until energy is needed. Then cells convert starch and glycogen into glucose, which is soluble, and respire the glucose.

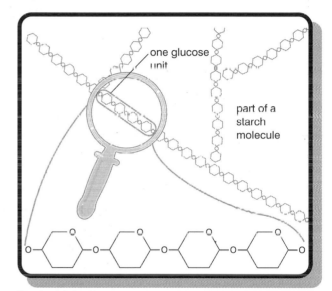

Structure of starch

15.3 Tests for carbohydrates

Test for glucose and other reducing sugars

Glucose is a reducing sugar. It will reduce **Benedict's solution**, and **Fehling's solution**, which both contain copper(II) sulphate, $CuSO_4$, and are, therefore, blue. The product is copper(I) oxide, Cu_2O, which is a red solid. When glucose is added to Benedict's solution and the mixture is kept in a warm water bath for several minutes, a red precipitate appears.

Fructose also tests positive. Of the disaccharides maltose and lactose give positive results, but sucrose does not.

Test for starch

When starch is added to a solution of **iodine** in aqueous potassium iodide, a blue-black colour forms. Sugars do not give a positive reaction.

15.4 Digestion of starch

During **digestion** starch is broken down into glucose. Being soluble, glucose can pass through the gut wall and travel in the blood stream to the parts of the body where it is needed. There it is oxidised in **cellular respiration** with the release of energy (see 15.1).

The digestion of starch is:

starch + water → glucose

$(C_6H_{10}O_5)_n(s) + nH_2O(l) \rightarrow nC_6H_{12}O_6(aq)$

This is an example of **hydrolysis**: a reaction in which molecules of a compound react with water and split up to form smaller molecules.

In the body, digestion of starch takes place rapidly at body temperature under the influence of the enzyme, **amylase**. This illustrates what powerful catalysts enzymes are. When the reaction is carried out in the laboratory, hydrochloric acid of concentration about 1 mol/l is added to an aqueous solution of starch and warmed to about 70 °C. Polysaccharides break down into disaccharides and then into the monosaccharide glucose.

Following the course of digestion of starch

The natural process of digestion can be limited in the laboratory by using a solution of the enzyme amylase, which occurs in saliva. The amylase solution is added to a 4% starch solution. The progress of the reaction from starch to glucose can be followed:

a) by testing for starch

On a white tile 15–20 drops of a solution of iodine in aqueous potassium iodide are spotted side by side. A glass rod is used to add a drop of the starch–amylase mixture to one drop of the iodine solution. A blue-black colour shows that starch is present. More drops of solution are tested at half-minute intervals over a period of time. The starch-iodine colour becomes fainter, showing that starch is being used up.

b) by testing for glucose

At the start of the reaction between starch and amylase, a sample of the starch–amylase mixture tested with Benedict's solution shows no precipitate. When the test is repeated after intervals of time, the solution changes from blue to cloudy green to cloudy brown and eventually to a colourless solution with a red precipitate, showing that glucose has been produced.

The products of digestion

The monosaccharides formed when disaccharides and polysaccharides are broken down (by hydrolysis, see 15.1) can be separated by thin layer or paper chromatography. The figure below shows a typical chromatogram.

G = glucose   **F** = fructose

S = sucrose   **M** = maltose

Products of hydrolysis are:

P1 = sucrose   **P2** = maltose

P3 = starch   **P4** = glycogen

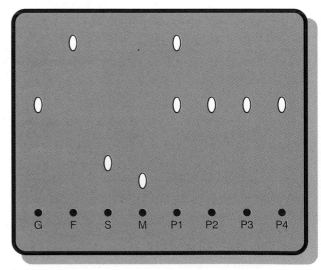

A chromatogram

You will see from the chromatogram that:

★ **Sucrose** is hydrolysed to a mixture of **glucose** and **fructose**. Sucrose is a disaccharide with molecules made of one glucose ring combined with a fructose ring.

★ **Maltose**, **starch** and **glycogen** are broken down (hydrolysed) to **glucose** only. Maltose is a disaccharide with molecules made of pairs of glucose rings. Starch is a polysaccharide with molecules which consist of chains of glucose rings.

15.5 Colloidal starch

Monosaccharides and disaccharides form true solutions. Molecules of starch are too large to form a true solution. Starch forms a **colloidal solution** in which molecules are suspended in and dispersed (spread) through the water. Starch cannot be separated from the solution by filtration. The molecules of starch are large enough to scatter light, as shown in the figure below. This effect is called the **Tyndall effect**, after its discoverer.

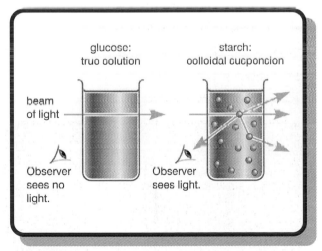

The scattering of light by a colloid

Dialysis

Dialysing membranes allow ions and small molecules to pass through them, but not large molecules or colloidal particles. The movement of ions and small molecules through dialysing membranes is called **dialysis**. Most animal membranes are dialysing membranes. Visking tubing is a dialysing membrane. Dialysis can be used to separate colloids, e.g. starch, from solutions that contain both colloidal particles and small molecules, e.g. glucose, as shown in the following diagram.

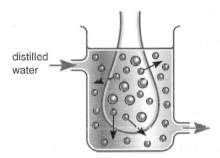

Small molecules, e.g. glucose, pass through the membrane, while colloidal particles, e.g. starch, cannot. The concentrations of small particles inside and outside the bag reach equilibrium.

With distilled water flowing slowly through the system, equilibrium cannot be established. Small particles pass out of the bag until only colloidal particles remain.

◉ = colloidal particle   ◉ = small molecule or ion

Dialysis

15.6 The alkanols

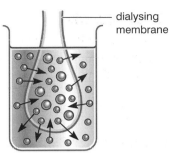

The **alkanols** are a homologous series of compounds with the general formula $C_nH_{2n+1}OH$. The members of the alkanol series all possess the **functional group** $\equiv$C—O—H, the hydroxyl group. They therefore have similar chemical reactions. Physical properties change gradually from one member to the next.

HOMOLOGOUS SERIES Page 40.

The alkanols	
name	**formula**
methanol	CH_3OH
ethanol	C_2H_5OH
propanol	C_3H_7OH
general formula	$C_nH_{2n+1}OH$

15

15.7 Ethanol

The drug

Ethanol (often called 'alcohol') is a drug, with a medical effect that is described as a **depressant** of the central nervous system. In small quantities, ethanol makes people feel relaxed. However, drinking large amounts of ethanol regularly causes damage to the liver, kidneys, arteries and brain. Methanol is so toxic that drinking only small amounts of it can lead to blindness and death.

Manufacture from fermentation

Ethanol is obtained by the fermentation of carbohydrates, such as glucose and starch.

$$\text{glucose} \xrightarrow{\text{enzyme in yeast}} \text{ethanol} + \text{carbon dioxide}$$

$$C_6H_{12}O_6(aq) \rightarrow 2C_2H_5OH(aq) + 2CO_2(g)$$

Yeast is added to fruit juices which contain sugars and left to ferment. Yeast contains enzymes. These are proteins which catalyse reactions in plants and animals (see page 18). Enzymes are affected by temperature and pH. The catalytic effect of the enzymes in yeast increases with temperature up to about 25 °C. Above this temperature, the reaction slows down because the enzymes are 'denatured': the three-dimensional structures change and the enzymes lose their catalytic activity. The optimum pH for the enzymes in yeast is about pH 7.

When the concentration of ethanol reaches 12–14%, it kills the yeast. More concentrated solutions of ethanol, e.g. whisky, are obtained by distillation.

Ethanol can be made from a number of starchy foods, such as potatoes, rice and hops. The starch is first treated with germinated barley (malt). This contains an enzyme that hydrolyses starch to a mixture of sugars. The sugars are then fermented by yeast.

Ethanol is sold as

- absolute alcohol (96% ethanol)
- industrial alcohol (methylated spirit), which has 5% methanol added to make the liquid unfit for drinking
- spirits (35% ethanol)
- beers and ciders (3–7% ethanol)
- wines (12–14% ethanol).

Oxidation

Ethanol is oxidised by air if certain micro-organisms are present. The reaction is used commercially to make ethanoic acid.

$$\text{ethanol} + \text{oxygen} \xrightarrow{\text{certain microorganisms}} \text{ethanoic acid} + \text{water}$$

$$C_2H_5OH(aq) + O_2(g) \rightarrow CH_3CO_2H(aq) + H_2O(l)$$

round-up

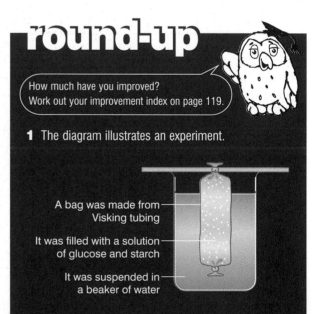

How much have you improved?
Work out your improvement index on page 119.

1 The diagram illustrates an experiment.

A bag was made from Visking tubing

It was filled with a solution of glucose and starch

It was suspended in a beaker of water

After 30 minutes, the liquid in the bag and the liquid in the beaker were tested.

liquid	Benedict's reagent	iodine solution
Visking tubing	red colour on heating	blue-black colour
Beaker	red colour on heating	no change

Explain these results. [1]

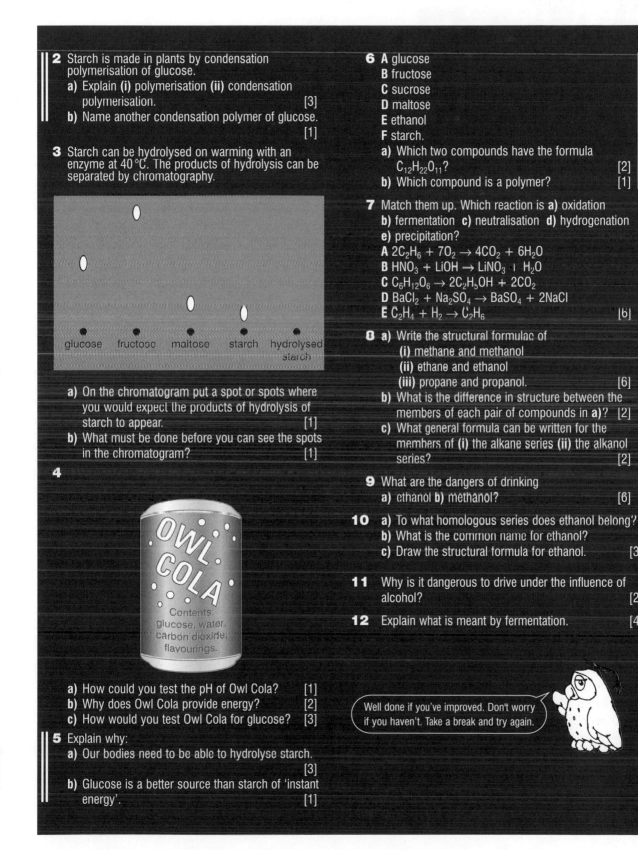

2 Starch is made in plants by condensation polymerisation of glucose.
  a) Explain (i) polymerisation (ii) condensation polymerisation. [3]
  b) Name another condensation polymer of glucose. [1]

3 Starch can be hydrolysed on warming with an enzyme at 40 °C. The products of hydrolysis can be separated by chromatography.

glucose   fructose   maltose   starch   hydrolysed starch

  a) On the chromatogram put a spot or spots where you would expect the products of hydrolysis of starch to appear. [1]
  b) What must be done before you can see the spots in the chromatogram? [1]

4

OWL COLA

Contents: glucose, water, carbon dioxide, flavourings.

  a) How could you test the pH of Owl Cola? [1]
  b) Why does Owl Cola provide energy? [2]
  c) How would you test Owl Cola for glucose? [3]

5 Explain why:
  a) Our bodies need to be able to hydrolyse starch. [3]
  b) Glucose is a better source than starch of 'instant energy'. [1]

6 A glucose
  B fructose
  C sucrose
  D maltose
  E ethanol
  F starch.
  a) Which two compounds have the formula $C_{12}H_{22}O_{11}$? [2]
  b) Which compound is a polymer? [1]

7 Match them up. Which reaction is a) oxidation b) fermentation c) neutralisation d) hydrogenation e) precipitation?
  A $2C_2H_6 + 7O_2 \rightarrow 4CO_2 + 6H_2O$
  B $HNO_3 + LiOH \rightarrow LiNO_3 + H_2O$
  C $C_6H_{12}O_6 \rightarrow 2C_2H_5OH + 2CO_2$
  D $BaCl_2 + Na_2SO_4 \rightarrow BaSO_4 + 2NaCl$
  E $C_2H_4 + H_2 \rightarrow C_2H_6$ [6]

8 a) Write the structural formulae of
     (i) methane and methanol
     (ii) ethane and ethanol
     (iii) propane and propanol. [6]
  b) What is the difference in structure between the members of each pair of compounds in a)? [2]
  c) What general formula can be written for the members of (i) the alkane series (ii) the alkanol series? [2]

9 What are the dangers of drinking
  a) ethanol b) methanol? [6]

10 a) To what homologous series does ethanol belong?
   b) What is the common name for ethanol?
   c) Draw the structural formula for ethanol. [3]

11 Why is it dangerous to drive under the influence of alcohol? [2]

12 Explain what is meant by fermentation. [4]

Well done if you've improved. Don't worry if you haven't. Take a break and try again.

107

Answers

1 Test yourself (page 10)

Matter and chemical reactions

1 solid (✓), liquid (✓), gas (✓).

2 A pure solid melts at a fixed temperature (✓); an impure solid melts over a range of temperature (✓).

3 Evaporation takes place over a range of temperature (✓). Boiling takes place at a certain temperature (✓).

4 Bubbles of vapour appear in the body of the liquid (✓).

5 A, C, E (✓✓✓)

6 table salt, water, sucrose (✓✓✓)

7 (i) Ni, S (✓✓) (ii) Ni, S, O (✓✓✓).

8 Crystals consist of a regular arrangement of particles (✓). As a result the surfaces are smooth (✓) and reflect light (✓).

9 The particles that make up the solid gain enough energy to break free from the attractive forces (✓) between particles which maintain the solid structure, and the particles move independently (✓).

Your score: ☐ out of 23

1 Round-up (page 15)

Matter and chemical reactions

1 At A, the temperature of the liquid is rising (✓). At B, the temperature stays constant because the liquid is boiling (✓) and all the heat is being used to convert liquid into gas (✓).

2 At C, the temperature of the solid is rising as it is heated (✓). At D the solid starts to melt (✓), and the temperature stays constant at the melting point (✓) while heat is used in the conversion of solid into liquid (✓). At E all the solid has melted (✓), and the temperature of the liquid rises as it is heated (✓). The sharp melting point shows that the solid is a pure substance (✓).

3
Compound	Elements present
Calcium oxide	calcium, oxygen (✓✓)
Magnesium chloride	magnesium, chlorine (✓✓)
Copper(II) sulphide	copper, sulphur (✓✓)
Copper(II) sulphate	copper, sulphur, oxygen (✓✓)
Potassium carbonate	potassium, carbon, oxygen (✓)

4 Most of a gas is space; the molecules are far apart (✓).

5 There is so much space between the molecules of a gas (✓) that it is easy for them to move closer together (✓) when the pressure is increased.

6 One example, e.g. heating a lump of bread dough, e.g. air in a hot air balloon expands and is less dense than the air outside the balloon (✓).

7 One example, e.g. increase in the pressure of air in car tyres, e.g. removing a dent from a table tennis ball by warming, e.g. a balloon filled with gas bursts if it is heated (✓).

8 The liquid vaporises (✓). A gas diffuses (✓) to occupy the whole of its container, i.e. the whole of the room (✓).

9 Liquids need energy to vaporise (see page 10) (✓). Aftershave lotion takes this energy from the skin (✓).

10 The appearance of the mixture (speckled yellow and grey) is different from that of the compound (dark grey throughout) (✓). The iron in the mixture reacts with dilute acids to give hydrogen (✓). The compound, iron(II) sulphide, reacts with acids to give hydrogen sulphide (✓), with a characteristic smell. The iron in the mixture is attracted to a magnet (✓).

11 45 (✓)

12 $CaCO_3(s) + 2HCl(aq) \rightarrow CaCl_2(aq) + CO_2(g) + H_2O(l)$ (✓✓✓ for state symbols, ✓ for factor 2)

13 $Fe^{2+}(aq) + 2OH^-(aq) \rightarrow Fe(OH)_2(s)$ (✓✓✓)

Your score: ☐ out of 43

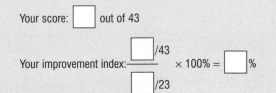

Your improvement index: $\dfrac{\boxed{}/43}{\boxed{}/23} \times 100\% = \boxed{}\%$

2 Test yourself (page 16)

Reaction speeds

1 Indigestion powders (✓) because the ratio surface area : mass is greater (✓).

2 a) Use smaller pieces of zinc (✓), use a more concentrated solution of acid (✓), raise the temperature (✓).

   b) In smaller particles, the ratio surface area : mass is larger (✓). At a higher concentration, collisions take place more frequently between hydrogen ions and zinc (✓). At a higher temperature, the hydrogen ions have higher energy (✓), and collide more frequently with the particles of zinc (✓).

3 An apparatus with a gas syringe, see page 17, and a clock (✓✓✓✓✓).

4 A catalyst is a substance which increases the rate of a chemical reaction without being used up in the reaction (✓✓).

5 Industrial manufacturers can make their product more rapidly or at a lower temperature with the use of a catalyst (✓✓).

6 Photosynthesis, photography (✓✓).

Your score: ☐ out of 20

2 Round-up (pages 18–19)

Reaction speeds

1 i) a (✓) **(ii)** f (✓)

2 a) F (✓) **b)** A + B, C + D, E + F (✓✓✓) **c)** C + E, D + F (✓✓)

3 A 2 (✓)   B 1 (✓)   C 3 (✓)   D 4 (✓)

4 a) Carbon dioxide is given off (✓).
b) (i) C   (ii) A   (iii) B (✓✓).
c) The ratio of surface area : volume differs (✓).

5 Use an apparatus such as that shown on page 17 (bottom) (✓). Using potato, measure the volume of oxygen formed at certain times after the start of the reaction (✓). Plot volume against time(✓). Repeat the measurement using manganese(IV) oxide. Compare the two graphs (✓).

6 You could take pieces of magnesium ribbon of the same length and therefore approximately the same mass (✓). You could find out how long it took (✓) for a piece of magnesium ribbon to react completely with a certain volume of acid (✓) of a certain concentration (✓) at different temperatures (✓). You could plot time against temperature or 1/time (rate) against temperature (✓).

7 a) Axes labelled correctly and units shown (✓), points plotted correctly (✓), points covering at least half of each scale (✓), smooth lines drawn through the points (✓).
b) B (✓)
c) Your line should should show very slow evolution of oxygen (✓).

Your score: ☐ out of 20

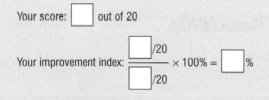

Your improvement index: ☐/20 ÷ ☐/20 × 100% = ☐ %

3 Test yourself (page 20)

Atoms and the periodic table

1 a) He, Ne, Ar, Kr, Xe (✓), a set of very unreactive gases present in air (✓).
b) Group 0 (✓)
c) (i) They have a full outer shell of electrons (✓).
(ii) They are very unreactive (✓). (Some take part in no chemical reactions. Krypton and xenon react with fluorine.)

2 Group 2 (✓)

3 Group 7 (✓)

4 Group 1 (✓)

5 a) Fluorine (✓), chlorine (✓), bromine (✓), iodine (✓).
b) Group 7 (✓)   **c)** Decreases (✓)
d) (i) NaF (✓), NaCl (✓), NaBr (✓), NaI (✓)
(ii) FeF_3 (✓), $FeCl_3$ (✓), $FeBr_3$ (✓), FeI_2 (✓)

6 An element in the block of the periodic table between Groups 2 and 3 (✓).
See page 23 (✓✓).

7 The number of protons = number of electrons (✓) and the positive charge on a proton has the same value as the negative charge on an electron (✓).

8 a) 19 (✓)   **b)** 20 (✓)

9 Their chemical reactions are identical because it is the electrons that determine the chemical behaviour (✓), and isotopes have the same electron arrangement (✓).

10 a) The number of protons (= number of electrons) in an atom of the element (✓).
b) The number of protons + neutrons in an atom of the element (✓✓).

11 a) $^{31}_{15}P$ (✓✓)   **b)** $^{39}_{19}K$ (✓✓)

12 Two in the first shell (✓), four in the second shell (✓).

Your score: ☐ out of 40

3 Round-up (pages 26–27)

Atoms and the periodic table

1 a) C (✓) **b)** Al (✓) **c)** He (✓) **d)** H (✓) **e)** K (✓)
f) F (✓)

2 a) E (✓) **b)** C (✓) **c)** B (✓) **d)** A (✓) **e)** E (✓)
f) D (✓)

3 a) elements (✓) **b)** atomic number (✓) **c)** number of outer electrons (✓) **d)** chemical properties/reactions (✓).

4 a) no. of positive particles = no. of negative particles (✓)
b) (i) nucleus (✓) **(ii)** space surrounding the nucleus (✓).

5 Ge was the undiscovered element below Si in Group 4 (✓✓). Mendeleev predicted that Ge would be similar to Si (✓).

6 First table He, 2, (2), B, 5, (2.3), Al, 13, (2, 8, 3), C, 6, (2.4), N, 7, (2.5), F, 9, (2.7), (18 ✓)
Second table a) 3, 4, 3   **b)** 3, 4, 2   **c)** 12, 12, 10
d) 9, 10, 10   **e)** 16, 16, 18, (15 ✓)

7 a) A and C (✓✓) **b)** B (✓) **c)** D (✓)

8 a) metallic (✓)   **b)** non-metallic (✓)   **c)** metallic (✓)
d) non-metallic (✓)

9 For example **(i)** Appearance; sulphur is dull whereas copper is shiny (✓). **(ii)** Sulphur is shattered by hammering, whereas copper can be hammered into shape (✓). **(iii)** Sulphur does not conduct heat or electricity, whereas copper is a good thermal and electrical conductor (✓). **(iv)** Sulphur is not sonorous; copper is sonorous (✓).

10 a) 27 (✓) **b)** 5 (✓) **c)** 0.7 (✓) **d)** 4 (✓)

11 a) $^{75}_{33}As$ (✓)   **b)** $^{235}_{92}U$ (✓), $^{238}_{92}U$ (✓), $^{239}_{92}U$ (✓)

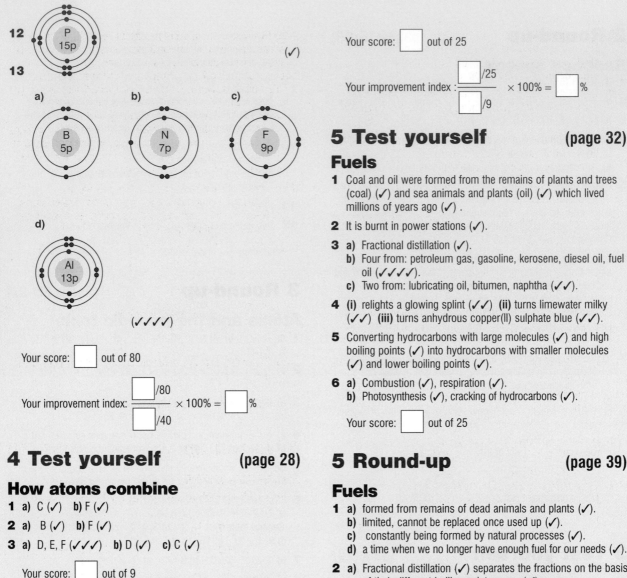

12

13

a) b) c)

d)

$(\checkmark\checkmark\checkmark\checkmark)$

Your score: ☐ out of 80

Your improvement index: $\dfrac{\boxed{}/80}{\boxed{}/40} \times 100\% = \boxed{}\%$

4 Test yourself (page 28)

How atoms combine
1 **a)** C ($\checkmark$) **b)** F ($\checkmark$)
2 **a)** B ($\checkmark$) **b)** F ($\checkmark$)
3 **a)** D, E, F ($\checkmark\checkmark\checkmark$) **b)** D ($\checkmark$) **c)** C ($\checkmark$)

Your score: ☐ out of 9

4 Round-up (page 31)

How atoms combine
1 H_2O see pages 30 and 31 ($\checkmark\checkmark$) HCl like HF see page 30 ($\checkmark\checkmark$), CO_2 see page 31 ($\checkmark\checkmark$), NCl_3 like NH_3 pages 30 and 31 ($\checkmark\checkmark$) CCl_4 like CH_4 pages 30 and 31 ($\checkmark\checkmark$), H_2S like H_2O pages 30 and 31 ($\checkmark\checkmark$), F_2 like Cl_2 page 29 ($\checkmark\checkmark$), CF_4 like CH_4 pages 30 and 31.
2 HBr similar to HF page 30 ($\checkmark\checkmark$), F_2 similar to Cl_2 page 30 ($\checkmark\checkmark$), H_2S similar to H_2O page 30 ($\checkmark\checkmark$).
3 See page 30 ($\checkmark\checkmark$).
4 An atom of argon has a full outer shell of 8 electrons ($\checkmark$).

Your score: ☐ out of 25

Your improvement index : $\dfrac{\boxed{}/25}{\boxed{}/9} \times 100\% = \boxed{}\%$

5 Test yourself (page 32)

Fuels
1 Coal and oil were formed from the remains of plants and trees (coal) ($\checkmark$) and sea animals and plants (oil) ($\checkmark$) which lived millions of years ago ($\checkmark$) .
2 It is burnt in power stations ($\checkmark$).
3 **a)** Fractional distillation ($\checkmark$).
  b) Four from: petroleum gas, gasoline, kerosene, diesel oil, fuel oil ($\checkmark\checkmark\checkmark\checkmark$).
  c) Two from: lubricating oil, bitumen, naphtha ($\checkmark\checkmark$).
4 **(i)** relights a glowing splint ($\checkmark\checkmark$) **(ii)** turns limewater milky ($\checkmark\checkmark$) **(iii)** turns anhydrous copper(II) sulphate blue ($\checkmark\checkmark$).
5 Converting hydrocarbons with large molecules ($\checkmark$) and high boiling points ($\checkmark$) into hydrocarbons with smaller molecules ($\checkmark$) and lower boiling points ($\checkmark$).
6 **a)** Combustion ($\checkmark$), respiration ($\checkmark$).
  b) Photosynthesis ($\checkmark$), cracking of hydrocarbons ($\checkmark$).

Your score: ☐ out of 25

5 Round-up (page 39)

Fuels
1 **a)** formed from remains of dead animals and plants ($\checkmark$).
  b) limited, cannot be replaced once used up ($\checkmark$).
  c) constantly being formed by natural processes ($\checkmark$).
  d) a time when we no longer have enough fuel for our needs ($\checkmark$).
2 **a)** Fractional distillation ($\checkmark$) separates the fractions on the basis of their different boiling point ranges ($\checkmark$).
  b) For uses see the diagram on page 33 ($\checkmark\checkmark\checkmark\checkmark\checkmark\checkmark$).
  c) Gasoline has a lower boiling point ($\checkmark$), lower ignition temperature ($\checkmark$) and lower viscosity than fuel oil ($\checkmark$).
  d) It does not vaporise ($\checkmark$) and does not ignite ($\checkmark$).
  e) Fuels from petroleum oil are important in transport ($\checkmark$), industry ($\checkmark$) and power generation ($\checkmark$). Oil is a source of valuable petrochemicals ($\checkmark$).
  f) Size of molecules increases from gasoline to fuel oil ($\checkmark$), therefore larger forces of attraction exist between molecules ($\checkmark$).
  g) **(i)** Increase ($\checkmark$) because larger molecules have larger forces of attraction between them ($\checkmark$) **(ii)** Decrease ($\checkmark$) because more difficult to separate the molecules and vaporise the fuels ($\checkmark$).
3 **a)** Poisonous ($\checkmark$); removed by soil bacteria ($\checkmark$).
  b) Causes respiratory difficulties ($\checkmark$); acid rain ($\checkmark$).
  c) Causes respiratory difficulties ($\checkmark$); washed out in rain ($\checkmark$).
4 **a)** From vehicle exhausts, power stations, factories ($\checkmark$).
  b) One of the causes of acid rain ($\checkmark$).
  c) Catalytic converters ($\checkmark$).

5 a) See page 38 (✓✓).
  b) See page 38 (✓✓).
  c) (i) Icecaps will melt (✓)   **(ii)** Flooding (✓)
    (iii) Decrease in wheat crop (✓).

6 Tall chimneys carry pollutants away from the area of the power station (✓). Taller chimneys do not prevent acid rain (✓).

7 a) Acidic water is released suddenly in the spring thaw (✓).
  b) Calcium hydroxide + sulphuric acid → calcium sulphate + water (✓✓)

$$Ca(OH)_2(s) + H_2SO_4(aq) \rightarrow CaSO_4(aq) + 2H_2O(l) \ (✓✓)$$

  c) Neutralisation (✓)

8 a) Iron + sulphuric acid → iron(II) sulphate + hydrogen (✓✓)
  b) Calcium carbonate + sulphuric acid → calcium sulphate + carbon dioxide + water

(✓✓✓)

  c) Calcium hydroxide + sulphuric acid → calcium sulphate + water (✓✓)

    (Nitric acid can be stated instead of sulphuric acid.)

9 a) Carbon dioxide (✓), carbon monoxide (✓), carbon (✓), water (✓).
  b) Oxides of nitrogen (✓), hydrocarbons (✓), sulphur dioxide (✓).

10 a) metal in the set between Groups 2 and 3 (✓) e.g. Pd, Pt, Rh (✓)
  b) $2CO(g) + 2NO(g) \rightarrow 2CO_2(g) + N_2(g)$ (4 ✓).

11 a) S in coal is oxidised to SO_2 (✓) **b)** contributes to bronchitis (✓) and to acid rain, page 53 (✓).
  c) 875 tonnes (✓✓).

12 i) plentiful (✓), pollutant combustion products (✓)
  ii) clean burning (✓), resources limited (✓).

13 a) tidal (✓) **b)** noisy (✓) **c)** detracts from scenery (✓)
  d) restricts passage of large ships (✓).

Your score: ☐ out of 84

Your improvement index: $\dfrac{\boxed{}/84}{\boxed{}/25} \times 100\% = \boxed{}\%$

6 Test yourself   (page 40)

Hydrocarbons

1 (✓), addition reactions (✓).

2 Alkenes are reactive (✓) and are therefore the starting materials for the manufacture of many other compounds (✓).

3 Bromine adds across the double bond to form $BrCH_2CH_2Br$ (✓).

4 a) (i) a compound of hydrogen and carbon only (✓) **(ii)** a hydrocarbon with only single bonds between carbon atoms (✓) **(iii)** a hydrocarbon with one or more carbon–carbon double bonds per molecule (✓) **b)** splitting molecules of alkanes (✓) into smaller molecules of different alkanes (✓) and alkenes (✓) and hydrogen (✓). Alkanes with smaller molecules are more useful as fuels than those with very large molecules (✓). Alkenes are used in the manufacture of many other compounds (✓). Hydrogen is used as a fuel (✓) and in the manufacture of ammonia (✓). With a catalyst cracking takes place at a lower temperature (✓), saving on fuel costs (✓).

5 a) Addition of hydrogen across a double bond (✓).
  b) Catalytic hydrogenation is used to convert unsaturated oils (vegetable oils) into saturated fats, which are solid, in the manufacture of margarine (✓).

6 a) propane C_3H_8 (✓)   (✓)

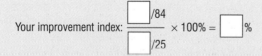

propene C_3H_6 (✓)   (✓)

  b) Propene decolourises bromine solution (✓); propane does not react (✓).

Your score: ☐ out of 26

6 Round-up   (page 42)

Hydrocarbons

1 a) (i) Ethane C_2H_6 (✓), ethene C_2H_4 (✓)
    (ii)

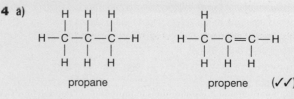

ethane   ethene

(✓✓)

  b) Ethane has only single bonds (✓); it is a saturated compound (✓). Ethene has a carbon–carbon double bond (✓); it is an unsaturated compound (✓).

2 a) Two from: hydrogen, bromine, water, sulphuric acid (✓✓).
  b) Addition reactions (✓).
  c) Hydrogen with a nickel catalyst (✓✓).

3 a) Carbon dioxide (✓) and water (✓).
  b) Alkenes are a source of valuable chemicals (✓).

4 a)

propane   propene   (✓✓)

b) Propene (✓)

c)

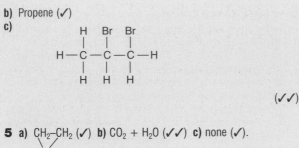

(✓✓)

5 a) CH₂–CH₂ (✓) b) $CO_2 + H_2O$ (✓✓) c) none (✓).

CH₂

d) propene $CH_3–CH=CH_2$ (✓✓)

6 a) C_2H_4 (✓) b) C_5H_{12} (✓) c) C_5H_{10} (✓) d) C_6H_{14} (✓).
e) C_8H_{16} (✓) f) $C_{12}H_{26}$ (✓)

7 (i) (✓✓)

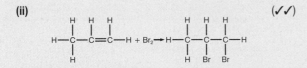

(ii) (✓✓)

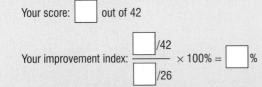

8 a) A set of compounds (✓) in which the formula of one member differs from the next by a –CH₂– group (✓) having similar chemical properties (✓) with a gradual change in physical properties from one member to the next (✓)
b) There are not enough H atoms in $C_{14}H_{30}$ to give 2 molecules of C_7H_{16} (✓).

Your score: ☐ out of 42

Your improvement index: $\dfrac{☐/42}{☐/26} \times 100\% = ☐\%$

7 Test yourself (page 43)

The ionic bond

1 Positive (✓), negative (✓), electrostatic (✓), crystal (✓).

2 Each atom of E loses two electrons to become E^{2+} (2.8) (✓).
Each atom of Q gains one electron to become Q^- (2.8.8) (✓).
The compound EQ_2 is formed (✓).

3 a) Magnesium bromide (✓) b) Iron(II) chloride (✓)
c) Iron(III) chloride (✓)   d) Sodium oxide (✓)
e) Barium sulphate (✓)

4 a) KBr (✓) b) $CaCO_3$ (✓) c) PbO (✓) d) $PbSO_4$ (✓)
e) AgCl (✓)

5 They are all 2, 8 (✓). O^{2-} (✓).

6 a) Cu atoms (✓) b) K^+ and Cl^- ions (✓✓)
c) Pb^{2+} and Br^- ions (✓✓).

7 a) Ions are in fixed positions in the solid (✓); free to move in solution (✓).
b) Ions are in fixed positions in the solid (✓); free to move in the liquid (✓).
c) This is a covalent compound (✓); it does not possess ions (✓).

8 CaF_2, KI (✓✓)

Your score: ☐ out of 32

9 A (✓) C (✓)

10 A (✓) B (✓) D (✓)

11 a) A compound which conducts electricity (✓) when molten (✓) or in solution (✓) and is split up in the process (✓).
b) An object which conducts electricity into or out of a cell (✓).

12 a) Cation – a positively charged atom or group of atoms (✓). Anion – a negatively charged atom or group of atoms (✓). For examples see table on page 47 (✓✓✓✓).
b) Ions are charged (✓) and move towards the electrode of opposite charge (✓).
c) In the solid the ions are held in place in a crystal structure (✓). In solution, they are free to move (✓).

7 Round-up (page 50)

The ionic bond

1 a) B (✓) b) D (✓) c) A (✓) d) C (✓) e) E (✓)

2 a) Copper(II) ions are blue (✓) and positively charged (✓). Permanganate ions are purple (✓) negatively charged (✓).

3 a) Copper (✓) at the negative electrode (cathode)(✓) and chlorine (✓) at the positive electrode (anode) (✓).
$Cu^{2+}(aq) + 2e^- \rightarrow Cu(s)$ (✓✓)
$2Cl^-(aq) \rightarrow Cl_2(g) + 2e^-$ (✓✓)
b) $MgCl_2$ consists of Mg^{2+} and Cl^- ions (✓) with strong forces of attraction between them (✓). CCl_4 consists of individual molecules (✓) with weak forces between them (✓).

4 S^{2-} (✓), Cl^- (✓), K^+ (✓), Ca^{2+} (✓)

5 On testing as shown the solutions give a) lead at cathode (✓)
b) hydrogen at cathode, chlorine at anode (✓✓) c) no change ✓.

6 AgOH, $Cu(OH)_2$, $Fe(OH)_2$, $Fe(OH)_3$, NH_4Br, $(NH_4)_2SO_4$, $NaNO_3$, $NaCO_3$, $AlCl_3$, $Al(OH)_3$, Al_2O_3, $Al_2(SO_4)_3$, $Ca(OH)_2$, CaO, $CaSO_4$ (15 ✓).

7 a) Sodium ions, Na^+ (✓) and chloride ions, Cl^- (✓).
b) Electrostatic attraction (✓).
c) A three-dimensional arrangement (✓) of alternate Na^+ ions and Cl^- ions (✓).

8 a) Calcium hydroxide (✓)   b) Sodium sulphite (✓)
c) Copper(II) carbonate (✓)   d) Magnesium hydrogencarbonate (✓)
e) Potassium nitrate (✓)

9 a) NH_4NO_3 (✓)　**b)** Na_2SO_4 (✓)　**c)** $(NH_4)_2SO_4$ (✓)
d) Al_2O_3 (✓)　**e)** $Zn(OH)_2$ (✓)

Your score: ☐ out of 59

Your improvement index: $\dfrac{\boxed{\ }/59}{\boxed{\ }/32} \times 100\% = \boxed{\ }\%$

8 Test yourself　(page 51)

Acids and bases

1 a) SA (✓)　**b)** WA (✓)　**c)** WB (✓)　**d)** WB (✓)
e) WA (✓)　**f)** N (✓)　**g)** SB (✓)

2 You could compare the rates (✓) at which hydrogen is given off (✓) in the reactions of the two acids with magnesium or another metal (✓),
or the rates (✓) at which carbon dioxide is given off (✓) in the reactions of the acids with a carbonate or hydrogencarbonate (✓).

3 a) Hydrochloric acid (✓)　**b)** For example, magnesium hydroxide (✓)　**c)** Citric acid (✓)　**d)** Ammonia (✓)

4 a) Sodium hydroxide converts grease into soap (✓).
b) Sodium hydroxide is a stronger base than ammonia (✓).
c) Wear glasses (✓) and rubber gloves (✓).
d) Sodium hydroxide would be too dangerous or corrosive (✓).
e) The saponification of fats by a weak base, e.g. ammonia, is very slow (✓).

5 a) 64 (✓)　**b)** 80 (✓)　**c)** 98 (✓)　**d)** 60 (✓)

6 a) 2.0 mol (✓)　**b)** 2 mol (✓)　**c)** 0.25 mol (✓).

7 a) 20.1 g (✓)　**b)** 24.5 g (✓)　**c)** 120 g (✓).

Your score: ☐ out of 33

8 Round-up

The mole　(page 55)

1 a) Gram, kilogram (✓)　**b)** Mole (✓)

2 Weigh out　**a)** the relative atomic mass in grams (✓)
b) the relative molecular mass in grams (✓).

3 a) 24 g (✓)　**b)** 69 g (✓)　**c)** 8 g (✓)　**d)** 16 g (✓)
e) 8 g (✓)　**f)** 16 g (✓)

4 a) 0.33 mol (✓)　**b)** 0.25 mol (✓)　**c)** 2.0 mol (✓)
d) 0.010 mol (✓)　**e)** 0.33 mol (✓)　**f)** 20 mol (✓)

5 a) 88 g (✓)　**b)** 980 g (✓)　**c)** 117 g (✓)　**d)** 37 g (✓)

6 a) 16 (✓)　**b)** 28 (✓)　**c)** 44 (✓)　**d)** 64 (✓)　**e)** 40 (✓)
f) 74.5 (✓)　**g)** 40 (✓)　**h)** 74 (✓)　**i)** 63 (✓)
j) 123.5 (✓)　**k)** 80 (✓)　**l)** 159.5 (✓)　**m)** 249.5 (✓)

Your score: ☐ out of 33

Solutions　(page 56)

1 $\text{Concentration of solution} = \dfrac{\text{amount of solute}}{\text{volume of solution}}$ (✓✓)

2 a) 0.2 mol/l (✓)　**b)** 0.02 mol/l (✓)
c) 0.2 mol/l (✓)　**d)** 8 mol/l (✓)

3 a) 0.25 mol (✓)　**b)** 0.01 mol (✓)　**c)** 0.05 mol (✓)
d) 0.0025 mol (✓).

Your score: ☐ out of 10

Acids and bases　(page 58)

1 $NH_3(g) + H_2O(l) \rightleftharpoons NH_4^+(aq) + OH^-(aq)$ (✓✓)
$CaO(s) + H_2O(l) \rightarrow Ca(OH)_2(aq)$ (✓✓)
$HCl(g) + aq \rightarrow H^+(aq) + Cl^-(aq)$ (✓✓).

2 a) more dilute (✓)　**b)** increases (✓)　**c)** less (✓)
d) decreases (✓).

3 a) greater than (✓)　**b)** greater than (✓)　**c)** equal to (✓).

4 i) A, C (✓✓)　**(ii)** D, F (✓✓).

5 A and C (✓✓)

Your score: ☐ out of 19

Your improvement index: $\dfrac{\boxed{\ }/19}{\boxed{\ }/33} \times 100\% = \boxed{\ }\%$

9 Test yourself (page 59)

Reactions of acids

1

hydrochloric acid + calcium oxide → calcium chloride + water (✓✓)

hydrochloric + sodium → sodium + carbon + water (✓✓✓)
acid   carbonate   chloride   dioxide

sulphuric acid + sodium hydroxide → sodium sulphate + water (✓✓)

sulphuric + copper(II) → copper(II) + carbon + water (✓✓✓)
acid   carbonate   sulphate   dioxide

sulphuric acid + magnesium → magnesium sulphate + hydrogen (✓✓)

nitric acid + calcium hydroxide → calcium nitrate + water (✓✓)

2

$Mg(s) + 2HCl(aq) \rightarrow H_2(g) + MgCl_2(aq)$ (✓)

$MgO(s) + H_2SO_4(aq) \rightarrow MgSO_4(aq) + H_2O(l)$ (✓✓)

$CaCO_3(s) + 2HCl(aq) \rightarrow CaCl_2(aq) + CO_2(g) + H_2O(l)$ (✓✓✓)

$2KOH(aq) + H_2SO_4(aq) \rightarrow K_2SO_4(aq) + 2H_2O(l)$ (✓✓)

3 a) see neutralisation page 62 (✓✓✓)
   b) see method 2 page 63 (✓✓✓).

4 a) $H^+(aq) + OH(aq) \rightarrow H_2O(l)$ (✓✓✓); spectator ions (✓)
   b) $CO_3^{2-}(s) + 2H^+(aq) \rightarrow CO_2(g) + H_2O(l)$ (✓✓✓).

5 Add solutions of strontium nitrate (✓) and sodium sulphate (✓) (or other soluble salts). Filter (✓). Dry (✓).

Your score: ☐ out of 39

9 Round-up (page 64)

Reactions of acids

1 a) A solid base can be separated by filtration (✓).
   b) If any acid remains it will be concentrated in the evaporation step (✓).
   c) **(i)** Note when the evolution of hydrogen stops (✓).
      (ii) Test with indicator paper to find out when the solution is no longer acidic (✓).
      (iii) Note when the evolution of carbon dioxide stops (✓).

2 a) Zinc + sulphuric acid → zinc sulphate + hydrogen (✓✓)
   b) Cobalt + sulphuric → cobalt + water (✓✓)
      oxide   acid   sulphate
   c) Nickel + hydrochloric → nickel + carbon + water
      carbonate   acid   chloride + dioxide
      (✓✓✓)
   d) Potassium + nitric → potassium + water (✓✓)
      hydroxide + acid   nitrate
   e) Ammonia + nitric acid → ammonium nitrate (✓✓)

3 a) Mix e.g. lead(II) nitrate solution and sodium sulphate solution (✓✓).
   b) Lead(II) + sodium → lead(II) + sodium (✓✓✓✓)
      nitrate + sulphate → sulphate + nitrate
      $Pb(NO_3)_2(aq) + Na_2SO_4(aq) \rightarrow PbSO_4(s) + 2NaNO_3(aq)$
      (✓✓✓✓)

      $Pb^{2+}(aq) + SO_4^{2-}(aq) \rightarrow PbSO_4(s)$ (✓✓✓)
   c) Do not touch the lead salts with your hands. Wash your hands after the experiment (✓).

4 a) Sodium sulphate + water (✓✓).
   b) Ammonium chloride, no other product (✓).
   c) Zinc chloride + hydrogen (✓✓).
   d) Copper(II) sulphate + water (✓✓).
   e) Calcium chloride + carbon dioxide + water (✓✓✓).

Your score: ☐ out of 40

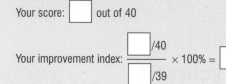

Your improvement index: $\dfrac{☐/40}{☐/39} \times 100\% = ☐\%$

Titration (page 66)

1 0.15 mol/l (✓)

2 a) 0.25 mol/l (✓)   **b)** 10.0 cm³ (✓)

3 a) Stoppo (✓)   **b)** Speed of action (✓); taste (✓).

4 Concentration of $H^+ = 1.0 \times 10^{-5}$ mol/l (✓✓)

Your score: ☐ out of 8

10 Test yourself (page 67)

Making electricity

1 Mg + Cu (✓).

2 a) Zn → Cu (✓)   **b)** decrease (✓).

3 a) **(i)** nr (✓)   **(iii)** nr (✓)   **(v)** nr (✓)
      (ii) magnesium(s) + zinc → zinc(s) + magnesium
             sulphate(aq)   sulphate(aq) (✓✓)
      (iv) copper(s) + silver → silver(s) + copper(II)
             nitrate(aq)   nitrate(aq) (✓✓)
      (vi) iron(s) + copper(II) → copper(s) + iron(II)
             sulphate(aq)   sulphate (aq) (✓✓)

   b) $Mg(s) + Zn^{2+}(aq) \rightarrow Mg^{2+}(aq) + Zn(s)$ (✓✓✓✓)

      $Cu(s) + 2Ag^+(aq) \rightarrow Cu^{2+}(aq) + 2Ag(s)$ (✓✓✓✓)

      $Fe(s) + Cu^{2+}(aq) \rightarrow Fe^{2+}(aq) + Cu(s)$ (✓✓✓✓)

4 a) Al oxidised (✓) Fe_2O_3 reduced (✓)
   b) Mg oxidised (✓) H^+ reduced (✓)
   c) Hg oxidised (✓) I_2 reduced (✓)
   d) SO_2 oxidised (✓) O_2 reduced (✓).

Your score: ☐ out of 32

10 Round-up (page 73)

Making electricity

1 a) from Al to Cu (✓)   **b)** Sugar is a non-electrolyte (✓)
   c) lower (✓).

2 a) $Br_2(aq) + 2e^- \rightarrow 2Br^-(aq)$ (✓✓)
   b) from B to A (✓)
   c) It allows ions to pass between the two beakers (✓).

3 **a)** copper (✓) **b)** The bigger the voltage, the further apart are the metals in the electrochemical series (✓).
  c) from the other metal to copper (✓) **d)** e.g. silver (✓)

4 **a)** Fe → Mn (✓) **b)** Fe^{2+} → Fe^{3+} (✓)
  c) **(i)** $2I^-$ → I_2 + $2e^-$ (✓✓) **(ii)** Fe^{3+} + e^- → Fe^{2+} (✓✓)
  d) $2I^-$ + $2Fe^{3+}$ → I_2 + $2Fe^{2+}$ (✓✓✓✓) **e)** I_2 → Fe (✓)

Your score: ☐ out of 22

Your improvement index: $\dfrac{\boxed{}/22}{\boxed{}/32} \times 100\%$ = ☐ %

11 Test yourself (page 74)

Metals and alloys

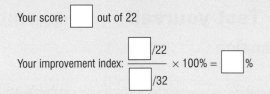

1 Metals can change shape without breaking (✓), conduct heat (✓), conduct electricity (✓).

2 Three from, for example, potassium, sodium, lithium, calcium, magnesium, aluminium, zinc, iron (✓✓✓).

3 Two from, for example, silver, gold, platinum (✓✓).

4 Three from, for example, lithium, sodium, potassium, calcium, magnesium (slowly) (✓✓✓). Products are hydrogen (✓) and the metal hydroxide (✓).

5 Hydrogen (✓) and the metal chloride (✓).

6 **a)** Group 1 (✓) **b)** Group 2 (✓)
  c) Between Group 2 and Group 3 (✓).

7 Electrolysis (✓) of the molten anhydrous chloride or oxide (✓).

8 The ore is heated with limestone and coke in a blast furnace (✓✓✓).

9 Two from, for example, aluminium, chromium, nickel (✓✓).

10 **a)** Na_2O (✓)
  b) HgO, Ag_2O (✓✓)
  c) PbO, CuO, ZnO (✓✓✓)
  d) Al_2O_3 (✓).

11 **a)** C (✓) **b)** A (✓) **c)** E (✓) **d)** D, F (✓✓).

12 **a)** E (✓) **b)** E (✓) **c)** D (✓) **d)** C (✓) **e)** A, C (✓✓).

13 **a)** $HgBr_2$ (✓) **b)** $CuCl_2$ (✓) **c)** Na_2O (✓)

14 **a)** 50% (✓) **b)** 40% (✓) **c)** 33% (✓)

15 1 g (✓)

Your score: ☐ out of 50

11 Round-up (page 81)

Formulae

1 D (✓)

2 **A** = Mg_3N_2 (✓), **B** = Fe_3O_4 (✓), **C** = SiO_2 (✓),
  D = $MgSO_4$ (✓), **E** = C_3H_8 (✓)

3 **a)** C = 80%, H = 20% (✓✓)
  b) S = 40%, O = 60% (✓✓)
  c) N = 35%, H = 5%, O = 60% (✓✓✓)
  d) Ca = 20%, Br = 80% (✓✓)

Your score: ☐ out of 15

Reacting masses (page 82)

1 8.0 g (✓)

2 71 g (✓)

3 0.05 g (✓)

4 3.06 kg (✓)

5 0.26 g (✓)

Your score: ☐ out of 5

Metals and alloys (pages 82–83)

1 **a)** $4Na(s) + O_2(g) \rightarrow 2Na_2O(s)$ (✓✓✓, one for each symbol or formula)
  b) $2Mg(s) + O_2(g) \rightarrow 2MgO(s)$ (✓✓✓)
  c) $2Zn(s) + O_2(g) \rightarrow 2ZnO(s)$ (✓✓✓)
  d) $3Fe(s) + 2O_2(g) \rightarrow Fe_3O_4(s)$ (✓✓✓)
  e) Similar to **b** (✓✓✓)
  f) Similar to **b** (✓✓✓)
  g) Similar to **b** (✓✓✓)

2 **a)** **(i)** $Mg(s) + 2HCl(aq) \rightarrow MgCl_2(aq) + H_2(g)$ (✓✓✓✓, one for each)
    (ii) $Fe(s) + 2HCl(aq) \rightarrow FeCl_2(aq) + H_2(g)$ (✓✓✓✓)
    (iii) $Sn(s) + 2HCl(aq) \rightarrow SnCl_2(s) + H_2(g)$ (✓✓✓✓)
  b) **(i)** $Mg(s) + H_2SO_4(aq) \rightarrow MgSO_4(aq) + H_2(g)$ (✓✓✓✓)
    (ii) $Fe(s) + H_2SO_4(aq) \rightarrow FeSO_4(aq) + H_2(g)$ (✓✓✓✓)

3 **a)** Magnesium sulphate and hydrogen (✓✓).
  e) Zinc sulphate and hydrogen (✓✓).
  b), **c)**, **d)** and **f)** No reaction (✓✓✓✓).

4 Copper alloys do not rust (✓). They are softer than iron (✓) and easier to mint (✓).

5 a) With water, calcium reacts steadily (✓), magnesium over several days (✓), iron rusts over a period of weeks (✓), and copper does not react (✓).

b) With dilute hydrochloric acid, calcium reacts extremely vigorously (✓), magnesium reacts in minutes (✓), iron reacts at moderate speed with warm acid (✓), and copper does not react (✓).

6 a) The blue colour of the solution fades (✓) and a reddish brown solid is precipitated (✓).
Zinc + copper(II) sulphate → copper + zinc sulphate (✓✓)
$Zn(s) + CuSO_4(aq) \rightarrow Cu(s) + ZnSO_4(aq)$ (✓✓)

b) Grey crystals appear (✓).
Zinc + lead(II) nitrate → lead + zinc nitrate (✓✓)
$Zn(s) + Pb(NO_3)_2(aq) \rightarrow Pb(s) + Zn(NO_3)_2(aq)$ (✓✓)

7 Z > X > Y (✓✓)

8 a) Au (✓) **b)** Na (✓), Mg (✓) **c)** Zn(✓), Fe (✓)
d) Na (✓), Mg (✓), Zn (✓), Fe (✓) **e)** Na (✓)
f) Mg (✓), Al (✓), Zn (✓), Fe (✓) (Na reacts with water instead of with Pb^{2+}).

9 A: electrolysis (✓) of the molten anhydrous (✓) chloride or oxide (✓).
B: reduction of the oxide (✓) with carbon or carbon monoxide (✓).
C: electrolysis of a solution of the chloride (✓) or reduction of the oxide (✓).
D: electrolysis (✓) of the molten anhydrous (✓) chloride or oxide (✓).

10 Rubidium (Group 1) reacts vigorously with cold water (✓), bursting into flame because of the hydrogen formed (✓) and forming a solution of the alkali rubidium hydroxide (✓).

11 For uses of aluminium related to properties, see table on page 79. (✓✓✓✓ for four uses, ✓✓✓✓ for four properties.)

12 Saving of Earth's resources (✓), saving of energy used in extracting the metal, (✓) limiting damage to the environment through mining (✓).

13 a) 6.6 g (✓) **b)** 3.6 litres at rtp (✓)

14 a) 1250 tonnes (✓) **b)** 9.4×10^8 litres (✓)

15 a) The hydroxides are strong alkalis (✓).

b) (i) potassium + water → hydrogen + potassium hydroxide (✓✓✓✓)
(ii) $2K(s) + 2H_2O(l) \rightarrow H_2(g) + 2KOH(aq)$ (✓✓✓✓)
c) H_2 e.g. 'pops' with lighted splint (✓); KOH gives the alkaline colour with an indicator (✓).

16 Zinc is lower in the electrochemical series than aluminium (✓). ZnO is much easier to reduce than Al_2O_3 (✓) for which electrolysis must be used (✓).

17 a) copper (✓)
b) (i) copper(II) oxide + hydrogen → copper + water (✓✓✓✓)
(ii) $CuO(s) + H_2(g) \rightarrow Cu(s) + H_2O(l)$ (✓✓✓✓)
c) (i) hydrogen (✓) **(ii)** copper(II) oxide (✓)
d) oxidation-reduction (✓)
e) e.g. PbO (✓)

18 a) A, D (✓✓) **b)** B (✓) **c)** E (✓) **d)** C (✓) **e)** F (✓).

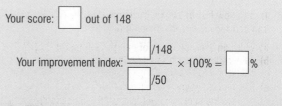

Your score: ☐ out of 148

Your improvement index: $\dfrac{☐/148}{☐/50} \times 100\% = ☐ \%$

12 Test yourself (page 84)

Corrosion

1 a) air, water, slight acidity (✓✓✓)
b) $Fe \rightarrow Fe^{2+} + 2e^-$ (✓) **c)** ferroxyl (✓)
d) turns blue with Fe^{2+} (✓).

2 Coating iron with zinc (✓). Zinc forms a protective layer (✓), excluding water and air (✓). If the layer is broken, zinc corrodes in preference to iron (✓).

3 A metal higher than iron in the electrochemical series (✓) is connected with an iron structure (✓). The other metal corrodes (✓) and protects the iron structure (✓). A metal such as magnesium or zinc is connected to the pipe (✓).

4 a) (i) Zn is higher than Fe in the electrochemical series (✓) therefore electrons flow from Zn to Fe (✓) and prevent $Fe \rightarrow Fe^{2+} + 2e^-$ (✓).
(ii) C is connected to a negative terminal (✓) therefore electrons flow into C (✓) and prevent $Fe \rightarrow Fe^{2+} + 2e^-$ (✓).
b) $Fe \rightarrow Fe^{2+} + 2e^-$ (✓).

Your score: ☐ out of 22

12 Round-up (page 88)

Corrosion

1 (1) Paint (✓) a barrier to water and air (✓) **(2)** Oil or grease (✓) barriers to water and air (✓) **(3)** Galvanised steel (✓) zinc is a barrier to water and air (✓) and also provides sacrificial protection (✓) **(4)** Connecting iron to the negative terminal of a battery (✓) makes electrons flow into the iron (✓) and prevents ionisation as Fe^{2+} (✓).

2 Connect iron to tin (✓) as shown in the diagram (✓).

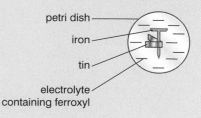

petri dish
iron
tin
electrolyte containing ferroxyl

Time the indicator turning blue (✓). Repeat with zinc in place of tin (✓).

3 Tin is below iron in the electrochemical series (✓). Iron corrodes in preference to tin (✓).

4 a) gives an even layer (✓) of the required thickness (✓).
b) Ni^{2+} cations are attracted to the negative electrode (✓).
c) $Ni^{2+}(aq) + 2e^- \rightarrow Ni(s)$ (✓✓✓)
d) $Ni(s) \rightarrow Ni^{2+}(aq) + 2e^-$ (✓✓✓)
e) It stays the same (✓). As fast as Ni^{2+} ions are discharged at the cathode, Ni atoms dissolve from the anode as Ni^{2+} ions (✓).

5 A, D and F (✓✓✓).

6 Nail B (✓). Copper is below iron in the electrochemical series (✓) therefore iron corrodes in preference to copper (✓), but zinc is above iron (✓) therefore zinc corrodes in preference to iron (✓).

7 a) galvanising **b)** paint **c)** sacrificial protection
d) galvanising **e)** stainless steel **f)** stainless steel
g) sacrificial protection **h)** chromium plate or plastic sheath
i) plastic sheath **j)** oil **k)** tin plate **l)** paint
m) chromium plate **n)** paint (14 ✓).

8 a) £47/year (✓✓) **b)** £30/year (✓✓).

9 Oil protects against rusting to some extent but not completely (✓). Contact with zinc protects completely against rusting (✓).

Your score: ☐ out of 56

Your improvement index: $\dfrac{\boxed{}/56}{\boxed{}/22} \times 100\% = \boxed{}\%$

13 Test yourself (page 89)

Plastics

1 oil (✓)

2 a) man-made (✓)
b) The supply of natural fibres is insufficient for our needs (✓).
c) (i) Cotton is absorbent and good for clothing (✓)
(ii) Nylon is strong and good for e.g. ropes (✓).

3 see page 90 for definitions (✓✓✓✓).

4 a) definitions page 89 (✓✓)
b) structure page 90 (✓✓)
c) uses page 91 (✓✓✓✓)

5 a) $CH_3-CH = CH_2$, propene (✓✓)
b)

(i) poly(chloroethene) (✓) **(ii)** PVC (✓) **(iii)** two of e.g. wellingtons, raincoats, insulation on electrical cable (✓✓).

6 Not broken down by micro-organisms in soil and air (✓). Plastics litter and plastics in landfill sites remain for many years (✓).

Your score: ☐ out of 26

13 Round-up (page 93)

Plastics

1 a) Less breakable (✓)
b) Cheaper (✓), less breakable (✓), non-toxic (✓)
c) Much less breakable (✓).
d)

2 a) When deformed, a plastic changes its shape (✓) and retains the new shape when the deforming force is removed (✓).
b) *Thermoplastic* – can be softened by heat (✓) many times (✓).
Thermosetting – can be moulded once only (✓).
Thermoplastic – individual chains can move with respect to one another (✓).
Thermosetting – chains are cross-linked (✓).

3 a) (i) (✓) **(ii)** (✓)
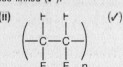

b) uses see page 91 (✓✓)

4 a) insulating, e.g. bakelite (✓✓)
b) resistant to heat, e.g. melamine (✓✓).

5 a) Oil – fractional distillation → fractions (✓). High b.p. fraction – cracking → products including ethene (✓). Ethene – polymerise → poly(ethene) (✓) **b)** millions of years (✓).
c) no (✓).

6 a) Not broken down by micro-organisms in soil and air (✓).
   b) Waste plastics remain in tips without decomposing for a long time (✓).
   c) The heat generated can be used (✓).
   d) carbon monoxide (✓)
   e) hydrogen chloride and chlorine (✓✓)
   f) recycle (✓).

Your score: ☐ out of 33

Your improvement index: $\dfrac{\boxed{}/33}{\boxed{}/26} \times 100\% = \boxed{}$ %

14 Test yourself (page 94)

Fertilisers

1 The roots of clover bear nodules (✓) containing nitrogen-fixing bacteria (✓) that convert nitrogen into compounds (✓).

2 They convert ammonium salts (which plants cannot use) (✓) into nitrates (which plants can use to make proteins) (✓).

3 They convert nitrates (✓) and ammonium salts (✓) into gaseous nitrogen (✓).

4 Nitrogen compounds (ammonium salts and nitrates) (✓), phosphates (✓) and potassium salts (✓).

5 Nitrates encourage the growth of algae (✓). When algae die and decay, they use up the dissolved oxygen (✓) and fish cannot live in the water (✓).

6 $(28/60) \times 100 = 47\%$ (✓✓)

7 High pressure (about 350 atm) (✓), a moderate temperature (about 450°C) (✓) and a catalyst (iron) (✓).

8 a)

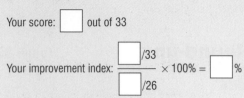

(✓✓✓)

   b) **(i)** penetrates to the roots of plants (✓)
       (ii) washed out of the soil (✓).

9 Phosphate rock (✓), sulphur (✓), potassium chloride (✓).

Your score: ☐ out of 27

14 Round-up (page 100)

Fertilisers

1 a) 82% (✓) **b)** 35% (✓) **c)** 21% (✓) **d)** 47% (✓).

2 Ammonia + hydrogen chloride → ammonium chloride (✓)

   $NH_3(g) + HCl(g) \rightarrow NH_4Cl(s)$ (✓✓)

3 $NH_4Cl(s) + NaOH(aq) \rightarrow NH_3(g) + NaCl(s) + H_2O(l)$ (✓✓✓)

4 **A** nitrogen (✓), **B** litmus or universal indicator (✓), **C** hydrogen chloride (✓),
   D alkali, e.g. sodium hydroxide (✓), **E** hydrogen (✓).

5 a) Proteins (✓)
   b) Legumes have nitrogen-fixing bacteria (✓) in nodules on their roots (✓).
   c) One of: by lightning; fixation by bacteria; Haber process; in vehicle engines (✓).
   d) Crops are harvested (✓), not left to decay and return nitrogen compounds to the soil (✓).
   e) Ammonium salts are converted by nitrifying bacteria (✓) into nitrates (✓), which plants can absorb (✓).

6 a) Slag (✓) **b)** Slag (✓), sylvite (✓) and *either* urea *or* ammonium sulphate (✓).

7 a) B (✓) **b)** C (✓) **c)** A (✓) **d)** E (✓) **e)** D (✓)

8 converting the element into compounds (✓), e.g. lightning storm (✓), e.g. Haber process (✓).

9 a) nitrogen and hydrogen (✓✓)
   b) 200 atm, 450°C, catalyst (✓✓✓)
   c) oxidation of NH_3 is exothermic (✓); combination of $N_2 + O_2$ is endothermic (✓).

Your score: ☐ out of 43

Your improvement index: $\dfrac{\boxed{}/43}{\boxed{}/27} \times 100\% = \boxed{}$ %

15 Test yourself (page 101)

Carbohydrates

1 a) $CO_2 + H_2O$ (✓✓) **b)** $CO_2 + H_2O$ (✓✓)
   c) $CO_2 + H_2O$ (✓✓) **d)** ethanol + CO_2 (✓✓)

2 a) C (✓) **b)** A (✓) **c)** E (✓) **d)** B (✓) **e)** D (✓)

3 a) D (✓) **b)** A (✓) **c)** C (✓) **d)** B (✓) **e)** A, E (✓✓)

4 a) With iodine (✓) starch → blue colour (✓);
   sucrose no reaction (✓)
   b) With Benedict's reagent (✓) glucose → orange-red (✓); sucrose no reaction (✓).
   c) Hydrolyse (✓); then sucrose → glucose (✓) which gives a positive with Benedict's reagent (✓) while maltose → fructose (✓) which gives a negative with Benedict's reagent (✓).

5 a) Burning more hydrocarbon fuels puts more carbon dioxide into the air (✓). Clearing forests removes trees which take carbon dioxide out of the air (✓).
   b) Enhanced greenhouse effect page 38 could lead to global warming (✓).

6 a) photosynthesis (✓)   **b)** carbon dioxide (✓)   **c)** oxygen (✓)
d) chlorophyll (✓)   **e)** starch (✓).

7 a) E (✓)   **b)** D (✓)   **c)** C (✓)   **d)** B (✓)

Your score: ☐ out of 43

15 Round-up   (pages 106–107)

Carbohydrates

1 There is a positive test for glucose both inside the Visking tubing and outside (✓). Some glucose has, therefore, passed through the Visking tubing into the beaker (✓). Iodine does not give a positive test for starch in the beaker, therefore, no starch has passed through the Visking tubing (✓). The reason for the difference is that glucose molecules are small (✓) and can pass through the pores in Visking tubing (✓), but the large starch molecules cannot (✓).

2 a) (i) page 90 (✓)   **(ii)** page 92 (✓✓)
b) e.g. cellulose (✓).

3 a) one spot only: glucose (✓)
b) use a locating agent (✓).

4 a) universal indicator (✓)   **b)** combustion of glucose in respiration (✓) is exothermic (✓).
c) Benedict's solution (✓), warm (✓), goes orange-red colour (✓).

5 a) Starch is insoluble (✓). It must be converted into glucose which is soluble (✓) and can be transported into blood stream to cells which need it (✓).
b) Glucose does not have to be hydrolysed before the cells can use it (✓).

6 a) C, D (✓✓)   **b)** F (✓)

7 a) B (✓)   **b)** A (✓)   **c)** D (✓)   **d)** C (✓)
e) F (✓)   **f)** E (✓)

8 a) (i)

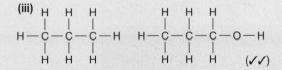

(ii)

(iii)

b) A hydroxyl group (✓✓).
c) (i) C_nH_{2n+2} (✓) **(ii)** $C_nH_{2n+1}OH$ (✓).

9 a) In the short term, slurred speech, blurred vision, long reaction times (✓✓); in the long term, damage to liver, kidneys, arteries and brain (✓✓).
b) Small amounts can cause blindness (✓) and death (✓).

10 a) The alcohols (✓)   **b)** 'Alcohol' (✓)
c)

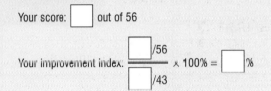

(✓)

11 Alcohol blurs the vision (✓) and increases reaction times (✓).

12 The action of enzymes (✓) on carbohydrates (✓) to form ethanol (✓) and carbon dioxide (✓).

Your score: ☐ out of 56

Your improvement index: $\dfrac{☐ /56}{☐ /43} \times 100\% = $ ☐ %

THE PERIODIC TABLE OF THE ELEMENTS

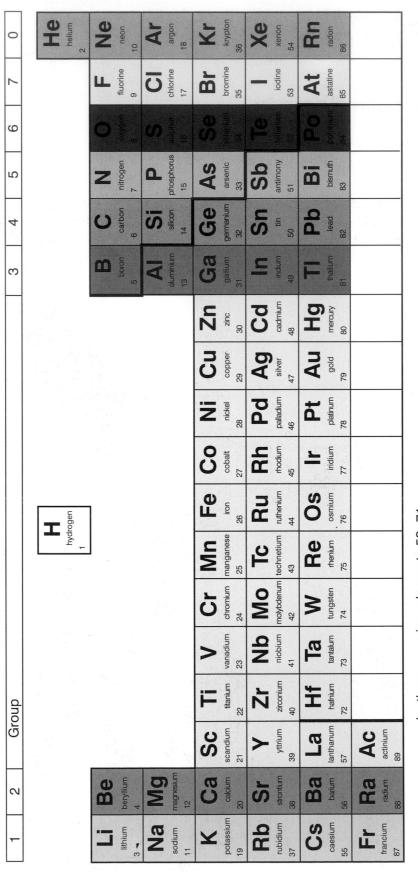

lanthanum series: elements 58–71
actinium series: elements 90–103

Mind Maps

Reaction speeds (Topic 2)

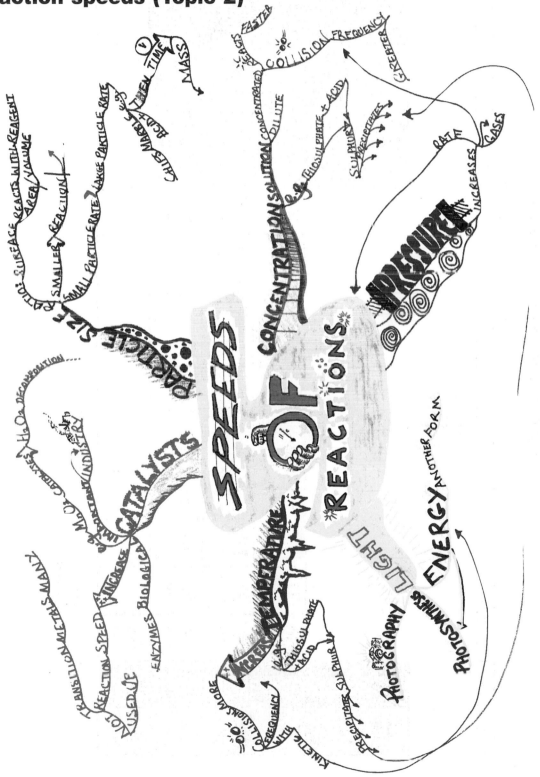

The structure of the atom (Topic 3)

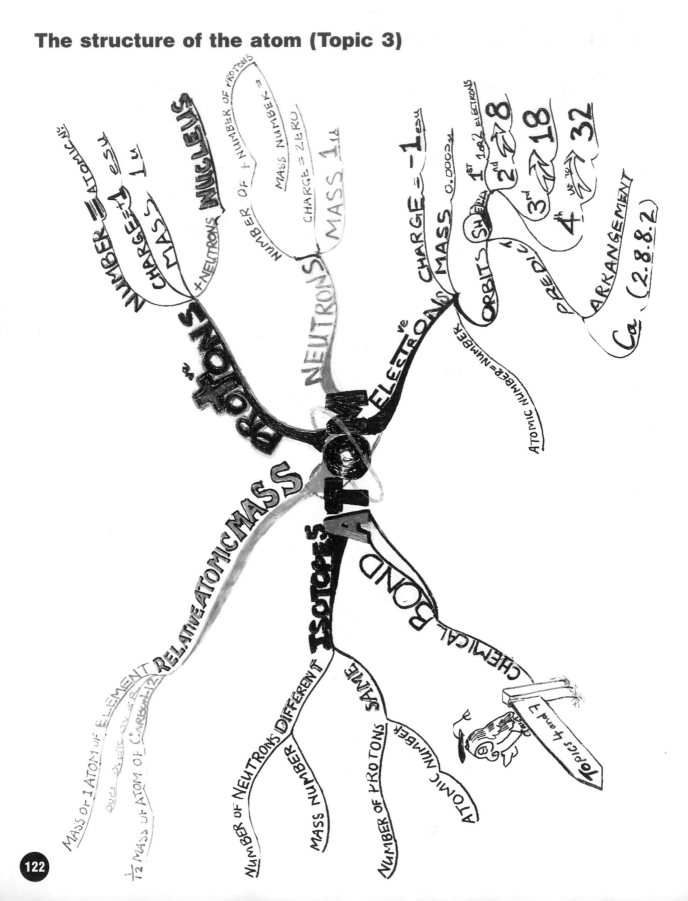

NUCLEUS

PROTONS
- NUMBER = ATOMIC Nº
- CHARGE = +1 esu
- MASS = 1 u

NUMBER OF PROTONS + NUMBER OF NEUTRONS

NEUTRONS
- MASS NUMBER = NUMBER OF PROTONS
- CHARGE = ZERO
- MASS 1 u

ATOM

ELECTRONS
- CHARGE = −1 esu
- MASS = 0.0005 u

ORBITS (SHELLS)
- 1st 1 or 2 ELECTRONS
- 2nd 8
- 3rd 18
- 4th up to 32

PREDICT

ATOMIC NUMBER

ARRANGEMENT
Ca (2.8.8.2)

RELATIVE ATOMIC MASS
- MASS OF 1 ATOM OF ELEMENT OVER 1/12 MASS OF ATOM OF CARBON-12

ISOTOPES
- NUMBER OF NEUTRONS DIFFERENT
- MASS NUMBER
- NUMBER OF PROTONS SAME
- ATOMIC NUMBER

CHEMICAL BOND
Topics 4 and 5

The chemical bond (Topics 4 and 7)

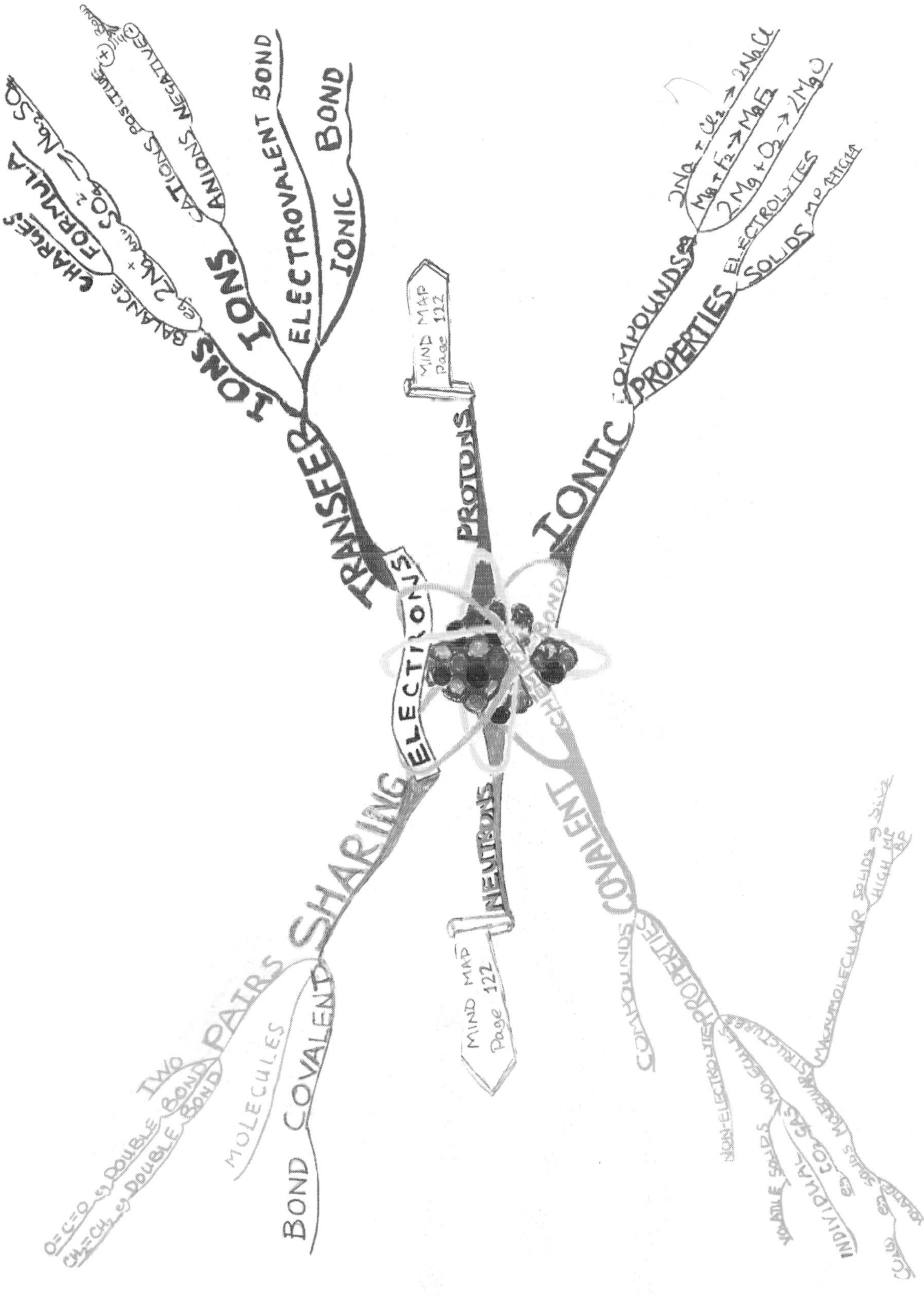

CHARGES FORMULA — Na_2SO_4

e.g. $2Na^+$ and SO_4^{2-}

BALANCE

FORMULA

IONS

IONS

CATIONS POSITIVE $(+)$

ANIONS NEGATIVE $(-)$

ELECTROVALENT BOND

IONIC BOND

TRANSFER

PROTONS

MIND MAP Page 122

ELECTRONS

NEUTRONS

MIND MAP Page 122

CHEMICAL BOND

IONIC

COMPOUNDS

$2Na + Cl_2 \rightarrow 2NaCl$

$Mg + F_2 \rightarrow MgF_2$

$2Mg + O_2 \rightarrow 2MgO$

PROPERTIES

ELECTROLYTES

SOLIDS HIGH M.P

SHARING

PAIRS

TWO

$O=C=O \rightarrow$ DOUBLE BOND

$CH_2=CH_2 \rightarrow$ DOUBLE BOND

MOLECULES

BOND COVALENT

COVALENT

PROPERTIES

NON-ELECTROLYTES

MOLECULAR STRUCTURES

INDIVIDUAL COVALENT

MACROMOLECULAR SOLIDS e.g. SiO₂ HIGH M.P

VOLATILE SOLIDS LOW M.P

Acids, bases and salts (Topics 8 and 9)

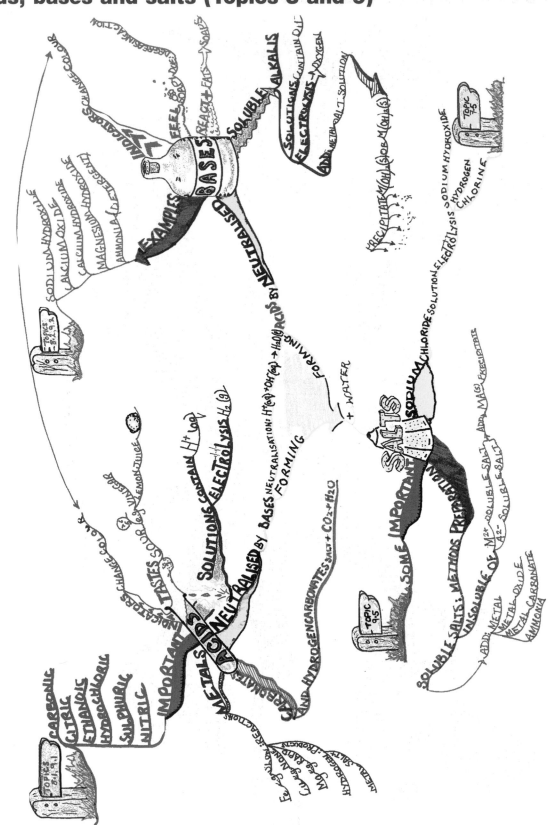

Metals and alloys (Topic 11)

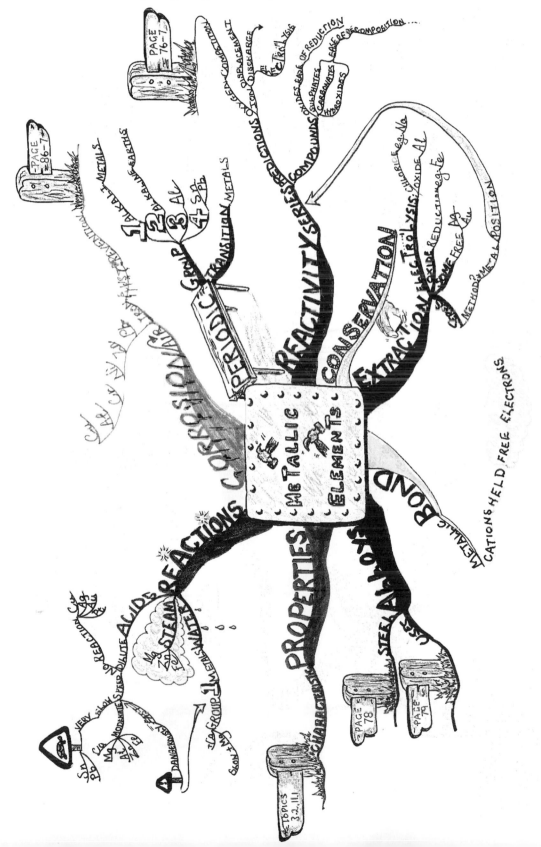

Alkenes and plastics (Topic 13)

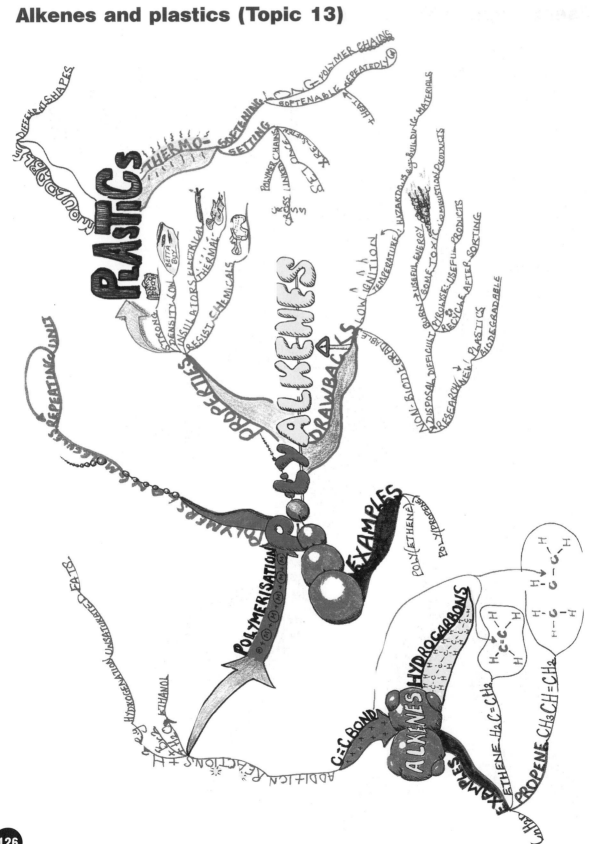

Fertilisers (Topic 14)

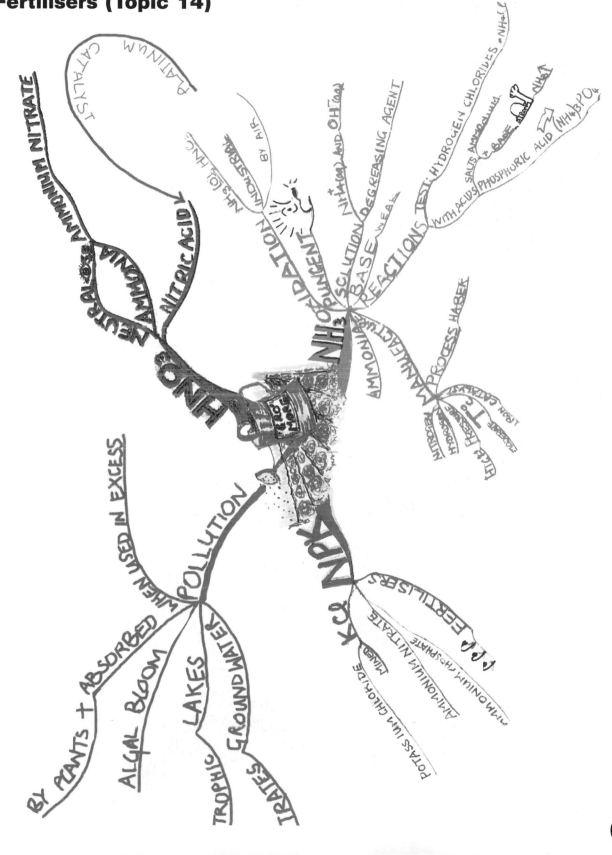

Index